Robotic Arm Control in 3D

Joshi S

Copyright © 2023 by Joshi S

All rights reserved. No part of this book may be reproduced in any manner whatsoever without written permission except in the case of brief quotations embodied in critical articles and reviews.

First Printing, 2023

Abstract

Control of flexible manipulators finds a lot of applications in the modern day world, especially in the field of avionics, robotics & the smart intelligent systems in the nano medical fields. Robot manipulators finds importance in this digital industrialized mechatronics age & are used in almost all the fields of applications ranging from product development to the medical field. Models of robot manipulators are important components of a robot motion control system. Advancement in the research of the robotic manipulators could be divided into 2 parts, viz., Rigid Manipulator (RM) and Flexible Manipulator (FM).

Most of Robotic Manipulators (RM) are designed with steel or aluminium frames for increased tensile strength. The result - rigid robotic arms, which are heavy and immobile. With advances in material technology, robotics and the increasing demand of light-weight and portable robotic arms, have been the catalyst for research in the design and control of flexible robotic manipulators, which are used nowadays for a large number of applications. Flexibility comes from the use of plastic, light weight (*Al*) or carbon-fiber frames, which significantly reduces the cost of manufacturing and the power consumption of the system.

The control algorithms of rigid manipulators is very easy, but when it comes to flexible manipulators it is very difficult because of the flexible nature of the plant. Thus, special algorithms must be developed to cater for flexibility in the robotic arms and these algorithms can be complex, considering that the system is non-linear and the control scheme employed to govern the plant to a pre-set location is needed. There are several control schemes to control the tip & the motors/actuators which are attached to the end of the flexible links. The current 21st century research is more inclined towards FM's because of their several advantages over the RM's because of the flexible nature in getting adaptable to the environment & to the sudden intermittent changes.

In this context, the research on control of flexible manipulators is being considered. Some of the prominent control schemes are enacted in this research work, which is used to regulate the motion of the flexible manipulators & take to a present location and also to track the set-points. In the work considered, a 1-link & 2-link flexible manipulator with 1-DOF & 2-DOF is being considered and the motors which are attached to the flexible links for actuation purposes are being controlled / regulated using pre-set values with less errors. Models are developed in the Simulink-Matlab environment & the simulation results observed, which shows the efficacy of the methodology developed.

In the yester years, after the rapid development of the industrial sector in the manufacturing world for the manufacturing of various products, rigid robots were being used. One drawback of this concept is - they are sturdy and being they operate in only one mode, i.e., the fundamental & the system is heavy (more mass, more inertia) as a result of which inertia will be more and more torque is needed to move from rest and to position it to a particular point using an appropriate control action.

Since most of the type of control techniques used in the work considered for controlling may need all the states for feedback, which may not be available for measurement, they may suffer from the problem of real time implementation and sometimes need a state observer for control purposes. These drawbacks could be rectified by the use of MROF (*a type of multi-rate output feedback controller, i.e., sampling @ 2 different rates, viz., τ & Δ*). The above mentioned drawbacks could be rectified by the use of MROF (POF, FOS, DSMC using output samples). An MROF based control technique can be applied to almost all the systems which are controllable and observable, while at the same time being simple enough as not to tax the computers too much.

State feedback algorithms can be converted into output feedback algorithms by the use of Multi-Rate Output Feedback (MROF) sampling. Consequently, the MROF based control strategies has the advantages of both the state feedback and output feedback control philosophies. Moreover, the MROF techniques does not need the system states and needs only the measurement of the system output for designing the controller. The resultant controllers would be output feedback based, thus being more practical than the more prevalent state feedback based approaches of controls.

This has motivated us to consider the problem of designing MROF based controllers (FOS, POF, DSMC) for tracking the set-point (position control) & for controlling the vibrations of SISO and MIMO systems (i.e., 1 & 2-link FMs). Control aspect becomes difficult once the mass comes into the picture as the inertia play will be there. More the mass, more is the force / torque needed to operate. It will be little difficult to adapt to the environment or change the shape of the rigid structure like that of a rubber-band, which is elastic in nature. Robot motion control is a key competence for robot manufacturers and current development is focused on increasing the robot performance, reducing the robot cost, weight, improving safety and introducing new functionalities.

Our main area of research is in control strategy, i.e., developing some novel hybrid robust algorithms for the control of flexible robots so that its usage is felt in the society, especially in the field of medicine for various applications such as cancer treatment, heart treatment, kidney ailments etc... The main advantage of flexible systems is that it can take the shape of the systems through which it can be sent like a nano-robot movement in a zig-zag rubber tube. This is the reason for the motivation to take up the research on the topic of robotic manipulators and this led us to the definition of the research problem, "*Control of n-link flexible robotic manipulators in 3D Euclidean space*", which was finalized as the research problem statement.

The research work that is undertaken was aimed to develop sophisticated control algorithms for control of flexible manipulator systems, where flexibility of the links and the joints play an important role, that too concentrating on the tip position accuracy & trajectory control of motors, which is our main desired objective. It has to be noted that the control of flexible manipulators eventually leads to the design and development of an automatic feedback control system so as to achieve a desired objective. The primary objectives of flexible robot arm control are accurate end-point positioning while a given task is being performed and robustness to any unmodelled dynamics.

The final result or the outcome or the end-result of the research work was aimed @ developing some efficient control algorithms which will accurately position the tip of the end-effector in spite of all non-linearities, noises, disturbances, vibrations, etc and to reduce the overall weight of the systems due to the flexible nature of the manipulator links, curb the vibrations / noises (unwanted signals) in just a couple of seconds using different types of sensors & actuators, adopt the hybrid type of control, i.e., position, velocity & vibration control along with motor tracking control. In short, to say, the outcome of the research is to show that when the flexible manipulator is placed with this developed robust controller, the flexible system will perform well and reaches the destination (output) in shorter lead times and will track the reference input.

The software tool that is used for the research work is Matlab with Simulink & various tool boxes such as control system tool box, signal processing tool box, control system tool box, optimization tool box, etc... In the following paragraphs, the various contributions of the research work are presented below as 4 different entities.

Contribution 1 : Control of 1-link & 2-link flexible robotic manipulator using PID control scheme for controlling the set-points (position).

Contribution 2 : Control of 1-link & 2-link flexible robotic manipulator using Periodic Output Feedback (POF) scheme for controlling the set-points (position).

Contribution 3 : Control of 1-link & 2-link flexible robotic manipulator using Fast Output Sampling Feedback (FOS) scheme for controlling the set-points (position).

Contribution 4 : Control of 1-link & 2-link flexible robotic manipulator using Discrete Sliding Mode (DSMC) with output sample scheme for controlling the set-points (position).

In the first case, the design, development of the simulink model & the control scheme to control / track the speed of the motor for a 1-link & 2-link flexible manipulator using the PID control strategy is considered. Controllers are being designed in the Matlab-Simulink environment for a single link & double link flexible robotic manipulator cases. The developed simulink model is used to regulate the speed of a DC electric motor which is connected to base of the single link flexible manipulator using one degrees-of-freedom PID control & 2-DOF PID control with the set-point weighting being set in the model. Note that one motor which is attached to the base of the manipulator which corresponds to one DOF. PID controller inside the blocks & the PID equations are converted into block-sets in the simulink diagram. Open loop & closed loop responses are also observed. Simulations are carried out with & without the controller to show the authenticity of the proposed control strategies in comparision with the work done by other researchers.

In the second case, the design of Periodic Output Feedback (POF) controllers for the developed SISO and MIMO models of the 1-link & 2-link flexible robotic manipulator is discussed. To design the controller, a state space model is obtained. 2 separate cases of the flexible robotic manipulator are presented here with, viz., the SISO (1-link) control, second-the MIMO (2-link) control. Open loop & closed loop responses are also observed. Simulations are carried out with & without the controller to show the authenticity of the proposed control strategies in comparision with the work done by other researchers.

In the third case, the design of Fast Output Sampling (FOS) feedback controllers for the developed SISO and MIMO models of the 1-link & 2-link flexible robotic manipulator is discussed. To design the controller, a state space model is obtained. 2 separate cases of the flexible robotic manipulator are presented here with, viz., the SISO (1-link) control, second-the MIMO (2-link) control. Open loop & closed loop responses are also observed. Simulations are carried out with & without the controller to show the authenticity of the proposed control strategies in comparision with the work done by other researchers.

In the fourth case, the design of Discrete Sliding Mode Controllers (DSMC) using the output samples for the developed SISO and MIMO models of the 1-link & 2-link flexible robotic manipulator is discussed. To design the controller, a state space model is obtained. 2 separate cases of the flexible robotic manipulator are presented here with, the SISO (1-link) control, second-the MIMO (2-link) control. A hybrid combination of the 2 different methods of control strategies has been used in this application, one, which uses a switching function in the control and results in quasi sliding mode motion, the other one does not use the switching in the control function. This strategy is used to demonstrate active vibration suppression in the 1 & 2-link flexible manipulators, i.e., to control the set-points & take it to a desired value in minimum time. Open loop & closed loop responses are also observed. Simulations are carried out with & without the controller to show the authenticity of the proposed control strategies in comparision with the work done by other researchers.

The outcome of the research work is to show that when the designed algorithm/s developed in the Matlab/Simulink environment is run, i.e., put in closed loop with the plant, the automatic tracking set-point is obtained with minimum computational time in comparison with the work done by the other researchers till date taking into consideration many of the drawbacks of the fellow researchers, thus enhancing and improving the performance of the existing algorithms with the end resulting in the following research outcome, "reaching the output settling value in minimum reasonable times".

Finally, the thesis has presented the investigations into control of a flexible manipulator system (1 & 2 link FMS). An automatic feedback control system to get the flexible system's output to the desired output was developed using the concepts of PID, POF, FOS & SMC methodologies. Note that to achieve this, dynamics & kinematics (forward kinematics & backward kinematics) of the robotic system was also developed for the flexible robot to achieve the target output. In a nutshell, research was conducted on the set-point control, end-point displacements, control of the joints of the tip-1 & tip-2 of the single link and double link flexible robotic manipulator using different control methods. CT state space model of the 1 & 2 link system was developed. 4 controllers were designed for the flexible manipulators to control the joint 1 of link 1 of 1-link manipulator & the joints 1 & 2 of – the 2 link flexible manipulator.

To conclude, a number of attempts was carried out in order to propose efficient & fast processing control algorithms for the control of single link & dual link flexible manipulators using different control methodologies, which are substantiated by the simulation results in Matlab/Simulink along with publications in journals & conferences.

Keywords : Robot, Flexibility, Matlab, Simulink, Control, POF, FOS, DSMC, PID, Response, Joint, Controller, Manipulator, Position, Actuator, Displacement, Discrete, Continuous, Sampling, Multirate, Sampling, Open loop, Closed loop, Response, SISO, MIMO, Link, Tracking, Model.

Table of Contents

Table of Contents
List of Figures
List of Tables
List of Nomenclature & Acronyms
List of Symbols

Chapter-1 : Introduction   **5**
1.1   A brief insight into the background research work   5
1.2   Specifications & models considered in the research work   11
1.3   Motivation / Problem statement definition   12
1.4   Objective of the research work   14
1.5   Outcome / Result of the research work   14
1.6   Scope of the research work   15
1.7   Methodology used in the research work   15
1.8   Software tool used in the research work   16
1.9   Contributions of the Ph.D. thesis   16
1.10   Flow of the thesis   17
1.11   References   19
1.12   List of publications   19

Chapter-2 : Literature Survey / Review   **20**
2.1   Control Schemes   20
2.2   Survey of the work done by various researchers till date   21
2.3   Drawbacks of the existing works   34
2.4   Conclusions   35

Chapter-3 : Controller of Flexible Manipulators using PID Controllers   **36**
3.1   Development of the simulink model for a single link flexible manipulator case with results & discussions   37
3.2   Development of the simulink model for a two link flexible manipulator case with results & discussions   45
3.3   Conclusions   51

Chapter-4 : Design of Controllers using Periodic Output Feedback – POF for 1 & 2-link FMs   **52**

4.1   A brief insight into the development of the Periodic Output Feedback (POF) Control Design for FMs   52
4.2   Control design for Single link flexible manipulator   57
4.3   Control design for dual link flexible manipulator   61
4.4   Development of the POF simulink model   64
4.5   Conclusions   65-67

Chapter-5 : Design of Controllers using Fast Output Sampling Feedback – FOS for 1 & 2-link FMs   **68**

5.1   A brief insight into the development of the Fast Output Sampling Feedback (FOS) Control Design for FMs   68
5.2   Control design for Single link flexible manipulator   76
5.3   Control design for dual link flexible manipulator   79
5.4   Development of the Simulink Model   85
5.5   Conclusions   87-88

Chapter-6 : Design of Controllers using Discrete Sliding Mode Control for 1 & 2-link FMs with output samples   **89**

6.1   A brief insight into the development of the DSM Control Design using o/p Samples for FMs   89
6.2   Control design for Single link flexible manipulator   98
6.3   Control design for dual link flexible manipulator   101
6.4   Development of the DSMC simulink model   105
6.5   Conclusions   106-107

Chapter-7 : Conclusions & scope for future work   **108**

7.1   Conclusions   108
7.2   Scope for future work   114

List of Figures

Fig. 1.1	: 1-link & 2-link flexible robotic manipulators supported at the base	11
Fig. 1.2	: Single link flexible manipulator attached to the base motor	12
Fig. 1.3	: 2 link FM attached to the base motor-link 1 & to the shoulder motor-link 2	12
Fig. 1.4	: Block diagram of the flexible manipulator control for 1-link and 2-link	16
Fig. 3.1	: Simulink model with one DOF PID control of a DC motor which is connected to the base of the flexible single link manipulator	38
Fig. 3.2	: Internal block diagram of the PID controller for a single link case	39
Fig. 3.3	: Tuning of the PID controller with the proportional, integral & derivative constants & PID controller tuning parameters KP, KI, KD values set in the block	39
Fig. 3.4	: Internal block diagram of the PID controller for a single link case	40
Fig. 3.5	: Setting up of the DC motor parameters in the function block parameter icon of the simulink model	40
Fig. 3.6	: Internal block-diagram of the DC motor	41
Fig. 3.7	: Result obtained after the tuning of the PID controller	42
Fig. 3.8	: Plot of step response for the reference tracking (tuned response & block response)	42
Fig. 3.9	: Set parameters & the tuned parameters	43
Fig. 3.10	: Output y angular displacement tracking the set-point (ref. i/p, r).....top simulation result along with the control input u .... Bottom simulation result	44
Fig. 3.11	: Output angular displacement of the motor (speed, ω rads/sec)	44
Fig. 3.12	: Torque input to the DC motor to which the link is connected	45
Fig. 3.13	: Set-point given as input to the PID controller to control the speed (angular displacement) of the DC motor	45
Fig. 3.14	: Control input u given as input to the PID controller	45
Fig. 3.15	: Developed simulink model of the tracking control of the 2-link flexible robotic manipulator using a single set-point	46
Fig. 3.16	: Internal block diagram of the PID controller for a two link case (for the base motor), similar one will be there for a shoulder motor	48
Fig. 3.17	: Internal block-diagram of the DC motor	49
Fig. 3.18	: Result obtained after the tuning of the PID controller	49
Fig. 3.19	: Output simulation results of the tracking control, control input & the set-point reference input for the base motor – 1 to which link 1 is connected (more control effort needed in this case)	51
Fig. 3.20	: Output simulation results of the tracking control, control input & the set-point reference input for the base motor – 2 to which link 2 is connected (less control effort needed in this case)	51
Fig. 4.1	: Graphical illustration of the POF control law developed	58
Fig. 4.2	: Open loop & closed loop response of a 1-link flexible manipulator	60

Fig. 4.3	: Bode diagram of a 1-link flexible manipulator to show effectiveness before & after the POF control	60
Fig. 4.4	: Open loop & closed loop response of a 2-link flexible manipulator for joint 1 at the base joint	63
Fig. 4.5	: Open loop & closed loop response of a 2-link flexible manipulator for joint 2 at the shoulder joint	63
Fig. 4.6	: Bode diagram of a 2-link flexible manipulator to show effectiveness before & after the POF control	64
Fig. 4.7	: Simulink model for design of a POF controller for a 1-link & 2-link flexible robotic manipulator system	65
Fig. 5.1	: Proposed flow-chart for the controller design using the FOS methodology	71
Fig. 5.2	: Graphical illustration of FOS feedback method	73
Fig. 5.3	: Block-diagrammatic representation of FOS feedback method of control strategy	73
Fig. 5.4	: Open loop & closed loop response of a 1-link flexible manipulator with SFB gain F & FOS gain L with control input u	78
Fig. 5.5	: Bode diagram of a 1-link flexible manipulator to show effectiveness before & after the FOS control	79
Fig. 5.6	: Open loop & closed loop response of a 2-link flexible robotic manipulator with SFB gain F1 & FOS gain L1 with control input u1, output displacement of tip-1, end of link-1 (shoulder joint), y1 & the zoomed version of the control i/p u.	82
Fig. 5.7	: Open loop & closed loop response of a 2-link flexible robotic manipulator with SFB gain F2 & FOS gain L2 with control input u2, output displacement of tip-2, end of link-2 y2	83
Fig. 5.8	: Bode diagram of a 2-link flexible manipulator system to show effectiveness before & after the FOS control	84
Fig. 5.9	: OL & CL response of the link-1 of a 2-link FRM with FOS gain L1 & control u1	84
Fig. 5.10	: OL & CL response of the link-2 of a 2-link FRM with FOS gain L2 & control u2	85
Fig. 5.11	: Simulink model for design of a FOS controller for a 1-link & 2-link flexible robotic manipulator system	85
Fig. 5.12	: Scope results of OL & CL for the base motor (link-1) ... amplitude more	86
Fig. 5.13	: Scope results of OL & CL for the shoulder motor (link-2) ...... amplitude less	86
Fig. 6.1	: Sliding mode control & the discrete sliding mode control	90
Fig. 6.2	: Proposed flow-chart for the controller design using the DSMC concepts	94
Fig. 6.3	: Bode diagram (frequency plot) for a single link-flexible robotic manipulator showing with & w/o control	100
Fig. 6.4	: System responses (plot of the system states), switching function & control input for a single link flexible robotic manipulator	100

Fig. 6.5	: Tip position control setting to the track point, equilibrium point	100
Fig. 6.6	: Control input plot showing the switching action	101
Fig. 6.7	: State trajectory (phase-plot) reaching equilibrium point of '0' showing tracking	101
Fig. 6.8	: State plots & the switching planes of the entire 2-link combined flexible system	103
Fig. 6.9	: Plot of the control effort u showing the switching action using the switching function s(k) for the single link flexible robotic manipulator	103
Fig. 6.10	: Plot of all the 4 system states for the link-1 of the flexible manipulator	104
Fig. 6.11	: Tip displacement of the link-1 of the flexible manipulator	104
Fig. 6.12	: Tip displacement of the link-2 of the flexible manipulator	105
Fig. 6.13	: Plot of the system states, switching function & control input required for the link-2 of the flexible manipulator	105
Fig. 6.13	: State trajectory (phase-plot) reaching equilibrium point of '0' showing tracking	106

List of Tables

Table 1.1 : Specs of the 1-link & 2-link flexible manipulator (length of link 2 < link 1) 11

Table 2.1 : A comparision of few of the authors works (advantages & lacunas) 34

Table 3.1 : Quantitative statistics of the control effort needed to control the motor 50

List of Nomenclature & Acronyms

Acronym	Meaning	Acronym	Meaning
#C1	Contribution 1	LQG	Linear Quadratic Regulator
#C2	Contribution 2	LQR	Linear Quadratic Regulator
#C3	Contribution 3	LTI	Linear Time Invariant
#C4	Contribution 4	MIMO	Multiple Input Multiple Output
3DE	3 Dimensional Euclidean Space	MPC	Model Predictive Control
AC	Alternating Current	MROF	Multi-Rate Output Feedback
AMM	Assumed Modes Method	NIT	National Institute of Technology
ANFIS	Adaptive Neuro-Fuzzy controller	OL	Open Loop
ANN	Artificial Neural Networks	OLIG	Open Loop Injection Gain
CL	Closed Loop	PD	Proportional Derivative
CLCS	Closed Loop Control System	PDE	Partial Derivative Equations
CNN	Convolution Neural Networks	Ph.D.	Doctorate of Philosophy
CT	Continuous Time	PI	Proportional Integral
DC	Direct Current	PID	Proportional Integral Derivative
DOF	Degree of Freedom	POF	Periodic Output Feedback
DSM	Discrete Sliding Mode	PVDF	Poly Vinyl Deryline Fluoride
DSMC	Discrete Sliding Mode Control	QFT	Quantitative Feedback Theory
DSP	Digital Signal Processing	QSMB	Quasi Sliding Mode Band
DT	Discrete Time	RGB	Red Green Blue
DTSMC	Discrete Time Sliding Mode Control	RM	Flexible Manipulator
		RM	Rigid Manipulator
FEM	Finite Element Method	RM	Robotic Manipulators
FJR	Flexible Joint Robot	ROI	Region Of Interest
FJR	Flexible Joint Robots	RT	Real Time
FLM	Flexible Link Manipulator	SCADA	Supervisory Control and Data Acquisition
FM	Flexible Manipulator		
FOS	Fast Output Sampling Feedback	SFB	State Feedback Gain
FOSSMC	Fast Output Sampling Sliding Mode Control Algorithm	SISO	Single Input Single Output
		SNR	Signal to Noise Ratio
FPGA	Field Programmable Gate Array	SOFB	Static Output Feedback
FRM	Flexible Robotic Manipulators	TF	Transfer Function
HDL	Hardware Descriptive Language	TI	Texas Instruments
IEEE	Institute of Electrical & Electronics Engineers	VHDL	VHSIC Hardware Description Language
IFOS	Integral Fast Output Sampling	VI	Virtual Instruments
IISc	Indian Institute of Science	VTU	Visvesvaraya Technological University
IIT	Indian Institute of Technology		
LMI	Linear Matrix Inequalities	w.r.t.	with respect to
LPM	Lumped Parameter Method	ZOH	Zero Order Hold
LQG	Linear Quadratic Gaussian		

List of Symbols

Symbol	Description	Symbol	Description
R^n	n-Dimensional Space	G	output injection gain matrix
l	Length	γ	controllability index
w	Width	A	system matrix
b	Thickness	B	input matrix
I	Moment of Inertia	C	output matrix
J	Moment of Inertia	D	transmission matrix
ρ	Density	x	state variable
E	Young's modulus	u	input variable
R_m	Tensile strength	y	output variable
ρ	Specific strength	Δ	sampling interval
E_ρ	Specific stiffness	$(\Phi_\tau, \Gamma_\tau, C)$	Δ (delta) system
n	No. of links	(Φ, Γ, C)	τ (tau) system
K_p	Proportional gain	k	time instants
K_i	Integral gain	N	sub-intervals
K_d	Derivative gain	$\mathbf{K}$	POF gain sequence
P	Proportional	$u(t)$	control input
I	Integral	$\rho(.)$	spectral radius.
D	Derivative	$y(\theta)$	displacement
$r(t)$	reference input	y_1 & y_2	displacement of links 1 & 2
b	Set point weight	$\mathbf{M}$	Mass matrix
c	Set point weight	$\mathbf{B}$	Damping matrix
N	No. of filter coefficients	$\mathbf{K}$	Stiffness matrix
s	complex variable	F	State Feedback Gain
R	Resistance	L	Fast Output Sampling Feedback gain
L	Inductance	u_1 & u_2	Control inputs to links 1 & 2
C	Capacitance	ρ_1, ρ_2 & ρ_3	performance matrices
H	Unit of inductance	s	sliding surface
F	Unit of capacitance	$\eta(k)$	Switch plane
Ω	Unit of resistance	q	DSMC parameter
K_m	Motor torque constant	ε	DSMC parameter
K_f	Field winding constant	x_1	State 1
J	Moment of Inertia	x_2	State 2
K_b	Back EMF constant	x_3	State 3
y	output control input	x_4	State 4
u	control input	$s(k)$	switching surfaces
r	reference set point	$x(0)$	initial condition
ω	Output angular displacement of the motor		
t	time		
τ	sampling time		

Chapter – 1

Introduction

In this chapter, a brief *introduction* about the background to the research topic chosen is presented along with its *applications*, i.e., a brief review of the concepts relating to the *control of flexible robotic manipulators* is presented to start with. The plant used for the control along with its *specs* is also presented. Also, the *motivation* obtained to take up the work on this excellent research topic is also presented along with how the problem has been defined leading to the *research problem statement definition*. The research plan outlay is also projected here as how we are solving the objective to arrive at the outcome. This is followed by the *main objectives*, the *outcome* & the *scope* of the research work, the advantages, etc. Finally, the *contributions*, i.e., the work done on the chosen research topic is presented in a nut-shell. The chapter concludes with the *flow of the interpretation* of the different chapters that are organized in the thesis.

1.1 A brief insight into the background research work

Control of flexible manipulators finds a lot of applications in the modern day world, especially in the field of avionics, robotics & the smart intelligent systems. Robot manipulators finds importance in this digital industrialized mechatronics age & are used in almost all the fields of applications ranging from product development to the medical field. Models of robot manipulators are important components of a robot motion control system. Advancement in the research of the robotic manipulators could be divided into 2 parts, viz., Rigid Manipulator (RM) and Flexible Manipulator (FM).

Most of Robotic Manipulators (RM) are designed with steel or aluminium frames for increased tensile strength. The result - rigid robotic arms, which are heavy and immobile. With advances in material technology, robotics, and the increasing demand of light-weight and portable robotic arms, have been the catalyst for research in the design and control of flexible robotic manipulators. Flexibility comes from the use of plastic, light weight (*Al*) or carbon-fiber frames, which significantly reduces the cost of manufacturing and the power consumption of the system. But with flexibility, a varying degree of inaccuracies in the robotic arm such as increased settling time, especially at the end point occurs.

The control algorithms of rigid manipulators are insufficient when it comes to flexible manipulators. Thus, special algorithms must be developed to cater for flexibility in the robotic arms, and these algorithms can be complex, considering that the system is non-linear. There are several control schemes to control the tip & the motors/actuators which are attached to the end of the links. To name a few of them are ….

- Fault tolerant control,
- Feed-forward control,
- Concurrent control,
- Programmable logic controls,
- Distributed control strategy,
- Positive position feedback control,
- Robust & non-linear control,
- Multi-rate output feedback controls,
- Static output feedback control,
- Pole placement techniques,
- PWM control strategy,
- Model predictive control (MPC),
- Linear quadratic regulator (LQG),
- Optimal feedback strategy,
- SCADA control,
- Supervisory control,
- Nonlinear Receding-Horizon Control,
- Fuzzy logic control,
- ANN & CNN control,
- Quantitative feedback theory (QFT) controls,
- PID control ….. *considered in the work*,
- Fast output sampling (FOS) feedback control….. *considered in the work*,
- Periodic output feedback (POF) control….. *considered in the work*,
- Discrete sliding mode control (DSMC) ….. *considered in the work*,

etc.

Each control scheme has got one advantage or the other & as such some of the researchers had used ***hybrid control strategies*** so that the dis-advantages of one are met by the other control strategy advantages. Some of the prominent control scheme is enacted in this research work, which is used to regulate the motion of the flexible manipulators & take to a present location and also to track the set-points. In the work considered, a 1-link & 2-link flexible manipulator with 1-DOF & 2-DOF is being considered and the motors which are attached to the flexible links for actuation purposes are being controlled / regulated using pre-set values with less errors. Models are developed in the Simulink-Matlab environment & the simulation results observed, which shows the efficacy of the methodology developed.

Since the dawn of robotics, the natural human being (living) has tried to develop different devices with the desire of creating artificial human beings (rigid robots & flexible robots – non living begins) similar to him/her which can help him/her in its daily work, carrying out repetitive or dangerous tasks. An important aspect of the human being, in contrast to the most of the living beings, is its ability to manipulate any object with the hands and have flexibility in doing any work, once the go signal is given. The desired objective will be met as the human being is capable of doing any type of manipulation in the 3DE space (R^n). Scientists have tried to provide to the robots this quality in the recent years, since the beginning of the robotics era by developing artificial hands.

Robot manipulators are an important research area in the modern day world used in almost all the fields of applications ranging from product development to the medical field. Models of robot manipulators are important components of a robot motion control system. Robotic manipulators could be divided into 2 types, viz.,

- Rigid Robotic Manipulator (RM) – Stationary and Mobile robots,
- Flexible Robotic Manipulator (FM) – Stationary and Mobile robots.

The current 21st century research is more inclined towards FM's because of their several advantages over the RM's because of the flexible nature in getting adaptable to the environment & to the sudden intermittent changes. In this context, the research on flexible manipulators is being considered. Some important *advantages* of the FM's are - light weight, low energy consumption, smaller size, more workspace, portability, economical, precise control, etc. Some of the limitations of flexible manipulator include - control complexity, non-minimum phase system, under actuation problem, non-collocation. The uncertainties include-truncation of flexible modes, control spillover, observation spillover, Eigenvalue problem, etc... The main reasons for the above complexities are the choice of dynamical models required good structural network and operation of FMs.

The dynamical model of a flexible manipulator depends on the type of modelling method & the control technology that is going to be used to control or regulate a particular parameter. The use of deformable or flexible robotic fingers (*in tasks where there is contact with the environment*) improves the limited capabilities of robotic rigid fingers. Elasticity of flexible fingers allows a greater adaptability between the manipulator and the object, moreover it avoids damages on the contact surfaces.

Active vibration control is one of the important problem to be tackled with in flexible RM's. One of the ways to tackle this vibration problem is to make the structure smart, intelligent, adaptive and self-controlling by making use of automatic feedback sensors. A precise mathematical model is required to start with in order to design a controller and then put in the feedback loop with the plant in order to control the vibrations when the plant (here-the actuator) is acted upon by an external force. Methods to model the flexible structures (FEM) with the help of smart materials using 2 types of modelling theories, viz., Euler-Bernoulli theory & the Timoshenko theory is gaining wide popularity nowadays.

Robotic manipulators are widely used to help in dangerous, monotonous, and tedious jobs. Most of the existing robotic manipulators are designed and build in a manner to maximize stiffness in an attempt to minimize the vibration of the end-effector to achieve a good position accuracy. This high stiffness is achieved by using heavy material and a bulky design not flexible. But, the existing heavy rigid manipulators are shown to be inefficient in terms of power consumption or speed with respect to the operating payload or settling times. Also, the operation of high precision robots is severely limited by their dynamic deflection, which persists for a period of time after a move is completed.

The settling time required for this residual vibration delays subsequent operations, thus conflicting with the demand of increased productivity. These conflicting requirements between high speed and high accuracy have rendered the robotic assembly task a challenging research problem. Also, many industrial manipulators face the problem of arm vibrations during high speed motion. In order to improve industrial productivity, it is required to reduce the weight of the arms and/or to increase their speed of operation.

For these purposes, it is very desirable to build flexible robotic manipulators. Compared to the conventional heavy and bulky robots, flexible link manipulators have the potential advantage of lower cost, larger work volume, higher operational speed, greater payload-to-manipulator-weight ratio, smaller actuators, lower energy consumption, better manoeuvrability, better transportability and safer operation due to reduced inertia. But the greatest disadvantage of these manipulators is the vibration problem due to low stiffness. For instance, it has been estimated that many cumulative hours would be spent in order to damp down the vibration to a small value in the remote manipulator system. Due to the importance and usefulness of these topics, researchers worldwide are nowadays engaged in the investigation of dynamics & control of flexible manipulators just like humans, whose body, links are flexible & can take to any position it wants in the 3D Euclidean space.

The control algorithms & the trajectory generation algorithms are 2 equally important components in the successful operation of any flexible robotic system. The problem of position control, velocity control, speed control, tip control, vibration control, robust control, and trajectory control for rigid robots is a well-known and completely understood issue, and rigid manipulators are extensively used in industries, but have some drawbacks. The desire for higher performance from the structure and mechanical specifications of robot manipulators has spurred designers to come up with Flexible Joint Robots known as the FJR's, because of its vast advantages over the rigid systems. Most robots have been designed to be mechanically stiff, because of the difficulty of controlling flexible members, not since rigidity, itself, is inherently attractive. On the other hand, several new applications such as space manipulators and articulated hands necessitate using FJRs.

In addition, as recently robots and humans have increasingly shared common spaces (*especially in the fields of medicine, biological robots and home automation*), it becomes necessary to consider the frequent physical contact between robots & humans. This also necessitates considering flexibility in manipulators as it leads to the development of systems with light weight & precise control. Out of these requirements, new control strategies have emerged, while traditional controllers used directly for FJRs had limited performance.

Since the 1980's after the invention of the IC chips, many attempts have been made to examine this problem, and, now, several methods have been developed world-wide, but have advantages as well as dis-advantages. The demands for increased robot accuracy coupled with high speed and large workspace requirements necessitate the evaluation of robot flexibility. The influence of flexibility on modelling and controller design must be better understood to achieve these requirements.

In the last couple of decades, many methods have been developed for modelling FMs. Generally, mathematical model of a manipulator is derived first, then the controller is designed, which takes care of the pre-set control action. Three methods are mainly used. These are finite element method (FEM), assumed modes method (AMM) and lumped parameter method (LPM). The most widely used method for modelling of FMs is AMM as the flexible system can operate in a large number of modes and the model selection can also be made.

As a result of which, the shape control has got more predominance now-a-days. It has several advantages like computational efficiency, flexibility in the choice of proper boundary conditions, etc. Also, coming to the control aspect, many people have worked on various vibration control issues such as the model predictive control, PID control, composite control, adaptive control, end-point control, joint control & mainly the vibration control in flexible systems. In the next para, some light is being thrown on the vibration suppression in flexible robots as vibration also affects the dynamical performance of FMs.

Vibrations also plays a very important role in the dynamical operation of any FM. Say, the flexible robotic system, when it is in operation, it is subjected to vibrations. These vibrations should not persist for a long amount of time, if so, the life span of the flexible system will be reduced. These vibrations should be suppressed as early as possible when the system is in operation. Hence, in recent years, the vibration control of structural characteristics using smart intelligent materials such as Piezoelectric, Shape Memory Alloys, Electro Rheological Fluids, Magneto Rheological Fluids, PVDF, Accelerometers, Optic fibres, Carbon nano-tubes, Pyro-electrics, Piezo-ceramics, etc..... has received considerable attention and has become an important problem in any type of flexible system.

One of the ways to tackle this vibration problem is to make the flexible structure smart, intelligent, adaptive and self-controlling by making use of intelligent materials as mentioned above; else they may affect its stability, longevity and its performance. These intelligent materials can be used to generate a secondary vibrational response in the FM system which has the potential to reduce the overall response of the system by the destructive interference with the original response of the system, caused by the primary source of vibration (operating due to noise or motor/actuator), thus saving the entire structure / plant from catastrophic disaster, breakdown & increasing the life span, reliability of the plant, the study being depicted as an important field in robotics, called as the '*flexible robotics or the flexi-bots*'. It is very much easy to control the rigid manipulators rather than the flexible robots because of the rigid nature of the links. Rigid manipulators when asked to go to a pre-set point goes & stops, but not the flexible bots because the links are not rigid.

1.2 Specifications & models considered in the research work

In the research work considered, 2 models are used to control the tip & the vibration of the motor connected to the joint. A pictorial representation of a 1-link & 2-link flexible robotic manipulators are shown in the Fig. 1.1 as

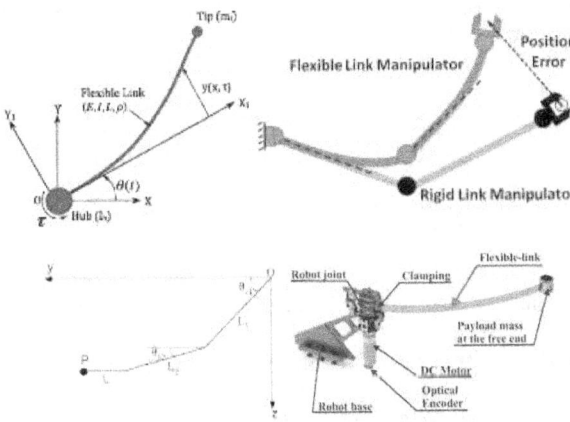

Fig. 1.1 : 1-link & 2-link flexible robotic manipulators supported at the base

Material properties	Symbol	Link-1	Link-2
Length	l	15 cm	10 cm
Width	w	1 cm	1 cm
Thickness	b	0.1 cm	0.1 cm
Moment of Inertia	I	0.3619E-10	0.3619E-10
Density kg/m^3	ρ	2.8	2.8
Young's modulus GPa	E	70	70
Tensile strength N/mm^2	R_m	180	180
Specific strength mm/kg	R_m/ρ	150	150
Specific stiffness Nm/kg	E/ρ	25	25

Table 1.1 : Specs of the 1-link & 2-link flexible manipulator (length of link 2 < link 1)

The model used in our work for the link-1 (one link flexible manipulator) & link-1 & link-2 (two link flexible manipulator) has the following specifications as shown in the Table 1.1 with its diagrammatic representation as shown in the Figs. 1.2 & 1.3 respectively. The material used is high grade low weight aircraft aluminium so that the overall weight/ratio is very less, density is good, speed of response is excellent & force at which it moves is also appropriate. A rectangular c/s aluminium link is considered for the research work.

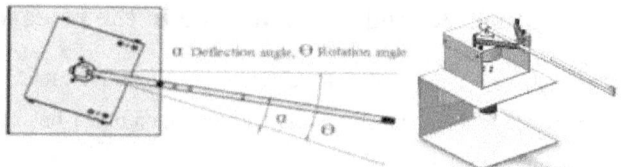

Fig. 1.2 : Single link flexible manipulator attached to the base motor

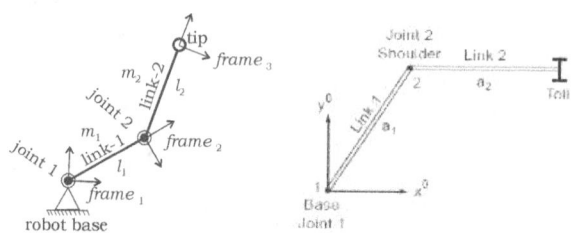

Fig. 1.3 : 2 link FM attached to the base motor-link 1 & to the shoulder motor-link 2

1.3 Motivation / Problem statement definition

The motivation for carrying out the research work is depicted in this section along with the problem statement of our research work.

In the recent years, with the development of technology at such a faster rate, the flexibility & adaptability concept has reached upto artificial human level also (as seen in the Hollywood movie – '*Terminator*', when the robot changes its shape once the reference signal is given & the desired objective is met). Finally, it has to be noted that for flexible joint models, the feedback control for path tracking has to be investigated by developing a perfect feedback control system. Our idea being to develop some hybrid robust control strategies to give such a reference point as the input in order to achieve the desired control objective. Hence, a modest attempt is going to be made in our research work by developing sophisticated control algorithms which has more advantages / benefits over the methods developed by the other researchers across the world.

In the yester years, after the rapid development of the industrial sector in the manufacturing world for the manufacturing of various products, rigid robots were being used. One drawback of this concept is - they are sturdy and being they operate in only one mode, i.e., the fundamental & the system is heavy (*more mass, more inertia*) as a result of which inertia will be more and more torque is needed to move from rest and to position it to a particular point.

Since most of the type of control techniques used for controlling may need all the states for feedback, which may not be available for measurement, they may suffer from the problem of real time implementation and sometimes need a state observer for control purposes. These drawbacks could be rectified by the use of MROF. The above mentioned drawbacks could be rectified by the use of MROF. An MROF based control technique can be applied to almost all the systems which are controllable and observable, while at the same time being simple enough as not to tax the computers too much.

State feedback algorithms can be converted into output feedback algorithms by the use of Multi-Rate Output Feedback (MROF) sampling. Consequently, the MROF based control strategies has the advantages of both the state feedback and output feedback control philosophies. Moreover, the MROF techniques does not need the system states and needs only the measurement of the system output for designing the controller. The resultant controllers would be output feedback based, thus being more practical than the more prevalent state feedback based approaches of controls.

This has ***motivated*** us to consider the problem of designing MROF based controllers (FOS, POF, DSMC) for tracking the set-point (position control) & for controlling the vibrations of SISO and MIMO systems (i.e., 1 & 2-link FMs). Control aspect becomes difficult once the mass comes into the picture as the inertia play will be there. More the mass, more is the force / torque needed to operate. It will be little difficult to adapt to the environment or change the shape of the rigid structure like that of a rubber-band, which is elastic in nature. Robot motion control is a key competence for robot manufacturers and current development is focused on increasing the robot performance, reducing the robot cost, improving safety, and introducing new functionalities.

Our main area of research is in control strategy, i.e., developing some novel hybrid robust algorithms for the control of flexible robots so that its usage is felt in the society, especially in the field of medicine for various applications such as cancer treatment, heart treatment, kidney ailments etc... The main advantage of flexible systems is that it can take the shape of the systems through which it can be sent like a nano-robot movement in a zig-zag rubber tube. This is the reason for the ***motivation*** to take up the research on the topic of robotic manipulators and this led us to the ***definition of the research problem***, "*Control of n-link flexible robotic manipulators in 3D Euclidean space*", which was finalized as the ***research problem statement***.

1.4 Objective of the research work

The research work that is undertaken by me under the guidance of my supervisor / guide is aimed to develop sophisticated control algorithms for control of flexible manipulator systems, where flexibility of the links and the joints play an important role, that too concentrating on the tip position accuracy, trajectory control of motors, which is our ***main desired objective***. This *objective* is going to be achieved using the short range / long range objective steps in a course of 5 years by developing sophisticated, efficient & low complexity control algorithms which are going to have a lot of advantages over the others.

It has to be noted that the control of flexible manipulators eventually leads to the design and development of an automatic feedback control system so as to achieve a ***desired objective***. The primary *objectives* of flexible robot arm control are accurate end-point positioning while a given task is being performed and robustness to any *unmodelled dynamics*. The amount of flexibility in the manipulator may be beneficial for some tasks, e.g., force generation in a bracing robot, and may be detrimental in others, e.g., rapid trajectory control, unknown payload positioning, i.e., controlling the speed of the motor attached to the link to tract a pre-set base value.

1.5 Outcome / Result of the research work

The ***final result*** or the ***outcome*** or the ***end-result*** of the research work was aimed @ developing some efficient control algorithms which will accurately position the tip of the end-effector in spite of all non-linearities, noises, disturbances, vibrations, etc… and to reduce the overall weight of the systems due to the flexible nature of the manipulator links, curb the vibrations / noises (unwanted signals) in just a couple of seconds using different types of sensors & actuators, adopt the hybrid type of control, i.e., position, velocity & vibration control along with motor tracking control. In short, to say, the *outcome* of the research is to show that when the flexible manipulator is placed with this developed robust controller, the flexible system will perform well and reaches the destination (output) in shorter lead times and will track the reference input.

The *outcome* of this research work has got wide application in the field of flexible industrial robotic manipulation, smart intelligent systems, space robotics, bio-medical engineering, structural health monitoring, flexible robotics, development of light weight systems, nano-robotics, etc… One of main outcome of our research work is to implement the work in some of the above mentioned areas and make it robust system so that in the event of any sensor / actuator / feedback mechanism failure also, the system will be come back to stability in no time, in the sense, the controller is going to be designed in such a way that the damping factor ξ is going to be between 0 and 1.

The outcome of the research work is to show that when the designed algorithm/s developed in the Matlab/Simulink environment is run, i.e., put in closed loop with the plant (motor), the automatic tracking set-point is obtained with minimum computational time in comparison with the work done by the other researchers till date taking into consideration many of the drawbacks of the fellow researchers, thus enhancing and improving the performance of the existing algorithms with the end resulting in the following research outcome, *"reaching the output settling value in minimum reasonable times"*.

1.6 Scope of the research work

The scope of the research work is presented in this section. The proposed research work has got a wide scope in industrial sectors, power plants, automobile, aerospace sector, turbines, centrifugal pumps, transportation, health care-medical field, computing systems, fans, antennas, motor control & in all the places where rotating or moving parts is present (dynamic machines).

1.7 Methodology used in the research work

The control strategy/methodology that is used in our research work is presented in this section. There are always constraints that make it difficult for programs to achieve optimal design, and each program team must weigh the trade-offs. These best practices provide an outline of the important considerations for the program design process that we have implemented. Hence, the methodology adopted in the present research work undertaken is depicted below with the help of a block-diagram as shown in the Fig. 1.4.

In fact, to say, we have developed an automatic feedback control system to get the flexible system's output to the desired output. Note that to achieve this, dynamics & kinematics (*forward kinematics & backward kinematics*) of the system has to be developed for the robot to achieve the target output. The output responses of the structure with & without the controller shall be observed, thus showing the control effectiveness and comparing of the research work done by us with other researchers in the past to establish the supremacy of the work developed.

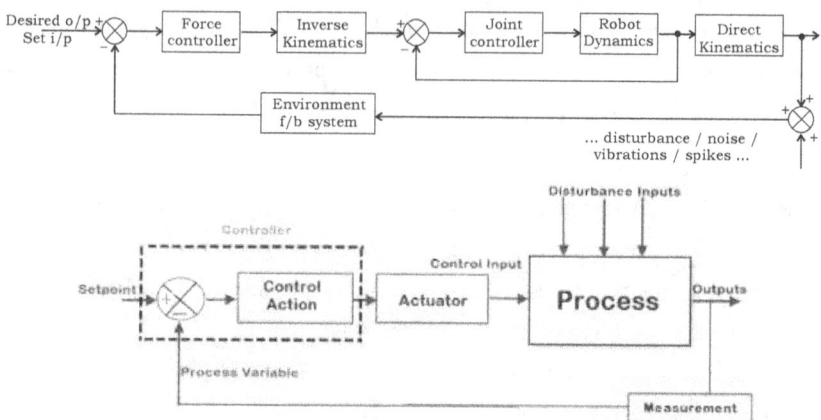

Fig. 1.4 : Block diagram of the flexible manipulator control for 1-link and 2-link
(*set-point tracking & vibration reduction @ the joint*)

1.8 Software tool used in the research work

The software tool that is used for the research work is Matlab 14 with Simulink & various tool boxes such as control system tool box, signal processing tool box, optimization tool box, etc...

1.9 Contributions of the Ph.D. thesis

In this section, the various contributions of the research work are presented below as 4 different entities from C1 to C4 done under the Matlab / Simulink environment, thus leading to 4 different contributory chapters.

#C1 - Contribution 1 : Control of 1-link & 2-link flexible robotic manipulator using PID control scheme for controlling the set-points (position) & the vibrations of the actuators.

#C2 - Contribution 2 : Control of 1-link & 2-link flexible robotic manipulator using Periodic Output Feedback (POF) scheme for controlling the set-points (position) & the vibrations of the actuators.

#C3 - Contribution 3 : Control of 1-link & 2-link flexible robotic manipulator using Fast Output Sampling Feedback (FOS) scheme for controlling the set-points (position) & the vibrations of the actuators.

#C4 - Contribution 4 : Control of 1-link & 2-link flexible robotic manipulator using Discrete Sliding Mode (DSMC) scheme for controlling the set-points (position) & the vibrations of the actuators.

1.10 Flow of the thesis

In this section, the flow of the Ph.D. thesis is presented in 7 different chapters one after the other as follows.....

Chapter 1 : presents a brief *introduction* about the background research work undertaken along with its stages of development and till the culminating stage with the problem statement, motivation, scope of the work, objectives ending with the organization of the thesis. This chapter also provides an introduction & assessment on the various parametric issues used for the control scheme.

Chapter-2 : gives an exhaustive summary of the literatures w.r.t background work in synchronization to the chosen research problem under different control schemes & its operating conditions. The related works presented here by various researchers across the globe is discussed in the form of exhaustive literature survey following the introductory chapter as '*Literature Survey*', which includes a number of articles w.r.t. the chosen research topic. It also includes the references of diverse technologies that has been used so far in the chosen field along with their drawbacks, lacunas, disadvantages & how we are going to address some of the issues in the considered work by us. It also includes the references of diverse technologies that has been used so far in the chosen field. Open loop & closed loop responses are also observed. This chapter also describes about the preceding available technologies which can be used to put in force the prevailing research till date along with the drawbacks & rectification along with a comparision table of the work done by various researchers till date.

Chapter 3 : This chapter presents the design & development of the simulink model & the control scheme to control / track the speed of the motor for a 1-link & 2-link flexible manipulator using the *PID* control strategy. Controllers are being designed in the Matlab-Simulink environment for a single link & double link flexible robotic manipulator cases. The developed simulink model is used to regulate the speed of a DC electric motor which is connected to base of the single link flexible manipulator using one degrees-of-freedom PID control & 2-DOF PID control with the set-point weighting being set in the model. Note that one motor which is attached to the base of the manipulator which corresponds to one DOF. PID controller inside the blocks & the PID equations are converted into block-sets in the simulink diagram. Open loop & closed loop responses are also observed. Simulations are carried out with & without the controller to show the authenticity of the proposed control strategies in comparision with the work done by other researchers.

Chapter-4 : In this chapter, the design of ***Periodic Output Feedback*** (POF) controllers for the developed SISO and MIMO models of the 1-link & 2-link flexible robotic manipulator is discussed. To design the controller, a state space model is obtained. 2 separate cases of the flexible robotic manipulator are presented here with, viz., the SISO (1-link) control, second-the MIMO (2-link) control. Open loop & closed loop responses are also observed. Simulations are carried out with & without the controller to show the authenticity of the proposed control strategies in comparision with the work done by other researchers.

Chapter 5 : In this chapter, the design of ***Fast Output Sampling*** (FOS) feedback controllers for the developed SISO and MIMO models of the 1-link & 2-link flexible robotic manipulator is discussed. To design the controller, a state space model is obtained. 2 separate cases of the flexible robotic manipulator are presented here with, viz., the SISO (1-link) control, second-the MIMO (2-link) control. Open loop & closed loop responses are also observed. Simulations are carried out with & without the controller to show the authenticity of the proposed control strategies in comparision with the work done by other researchers.

Chapter-6 : In this chapter, the design of ***Discrete Sliding Mode Controllers*** (DSMC) for the developed SISO and MIMO models of the 1-link & 2-link flexible robotic manipulator is discussed. To design the controller, a state space model is obtained. 2 separate cases of the flexible robotic manipulator are presented here with, the SISO (1-link) control, second-the MIMO (2-link) control. A hybrid combination of the 2 different methods of control strategies has been used in this application, one, which uses a switching function in the control and results in quasi sliding mode motion, the other one does not use the switching in the control function. These 2 applications are used to demonstrate active vibration suppression in the 1 & 2-link flexible manipulators, i.e., to control the set-points & take it to a desired value in minimum time. Open loop & closed loop responses are also observed. Simulations are carried out with & without the controller to show the authenticity of the proposed control strategies in comparision with the work done by other researchers.

Chapter 7 : concludes the research work done in the form of brief conclusions. The 4 contributory works (#C1 *to* #C4 – *PID, POF, FOS, SMC methods*) are compared with each other & with the work done by other authors to show the effectiveness of the methods proposed in this research work. The chapter also discusses about the scope for future work in this exciting field (*possible ways this work can be extended by future researchers*).

Chapter – 2

Literature Survey / Review

In the modern day technological world, automation plays a very important role in the human life ranging from domestic applications to the industrial applications. This automation makes use of various technological devices such as machines, computers & its accessories, etc… which could be used by the humans for various applications. Control of any device for a specific application is a very important aspect in the stability of the device and to make it work satisfactorily without any error to reach it to its set-point tracked value. During this tracking to the desired value, the system vibrates because of the moving parts involved, which has to be curbed down, else the performance would be deteriorated. One such device is the flexible robotic manipulator which could be used for a host of domestic applications to health sector to the industrial applications. A brief insight into the same was presented in the introductory chapter. In this chapter, a brief overview of the work done in the relevant field was conducted on the chosen research topic, *"Control of n-link flexible robotic manipulators in 3D Euclidean space"* and the same is presented in the form of an exhaustive literature survey [1] – [100] in the following paragraphs.

2.1 Control Schemes

There are several control schemes to control the flexibility of the robotic manipulators, such as modal reference adaptive control, self-tuning control, positive position feedback control and PID control, which are used to regulate the motion of the flexible manipulators. The control strategies for flexible manipulator systems can be classified as feed-forward (open loop) or feed-back (closed loop) control schemes, etc… The control schemes applied to flexible robots include proportional derivative control, computed torque control, active damping control, adaptive control, neural network based control, lead-lag control, sliding mode control, periodic output feedback control, fast output sampling feedback control, PWM control, hybrid control, neural & fuzzy control, stable inversion in the frequency domain, stable inversion in the time domain, algebraic control, optimal and robust control, input shaping control, passivity based control, and the boundary control, etc…. In all these schemes an efficient and accurate mathematical model is necessary to start with in order to develop the control scheme & obtain the simulated results in software-oriented approach.

2.2 Survey of the work done by various researchers till date

A number of researchers / authors have worked on the control of flexible manipulators (single link, double link, multi-link, n-link mechanisms) till date in various capacities and in all round aspects, starting from modelling, analysis, simulation upto the implementation level. To start with, 100s of research papers were collected from various sources, studied @ length & breadth and a review paper was published by us to start with in the field of the control aspects related to the research work undertaken. Rest of it as follows in the review.

Desoyer *et.al.* [1] compared the various modelling methods for light weight robots and discussed the effects of flexibility on possible control strategies. They examined the kineto-static method, the vibrational mode approach and the finite element method as a means of modelling flexible systems. **Troch and Kopacek** [2] discussed control strategies for flexible robots, designs based on model simplification and the effects of actuator dynamics. **Peng and Liou** [3] did experimental studies involving flexible mechanisms from a designer's point of view. They examined the identification of damping and mode shapes, vibration reduction and various means of measuring flexible mechanism responses.

Book & Kwon [4] described the modelling of flexibility, the large motion equations used and the design of flexible arms. He also presented the trajectory planning and some planning strategies for the control of flexible robot arms. **Cannon and Schmitz** [5] initiated the experiment to control the end-effector of a flexible manipulator by measuring the tip position and using that measurement as a basis for applying torque to the other end (joint) of the robot. However, they only considered only a linearized model and also the arm can sweep only in the XY plane (drawback), so that it is not affected by the gravity. Since then, many new control strategies are developed to control the flexible link vibration by using the strategy given by Cannon & Schmitz.

Recently, **Benosman and Vey** [6] presented a survey on the control of flexible manipulators which dealt with the multi-link manipulators and mainly the research works between 1990 and 2015 were cited and showed the developments in the 25 year period in the field of flexible robotics & its applications in all sectors. Many investigators worked to control the position of the end-effector of the single-link manipulators. Here, the methods used by different investigators are described in a nut-shell so that this article provided a base for many of the researchers who wanted to pursue their career in flexible robots.

While the computed torque control is used by **Thomas Looke** *et.al.* [7], an inversion based control schemes were used by **Singh and Schy** [8] for the end-point control of single-link flexible manipulator. **Inna Sharf** [9] used an active damping control scheme for a short-reach robot. Adaptive control schemes were used by **Menq and Chen** [10] for the tip-position control of single-link manipulators. **Korolov & Chen** [11] used robust control schemes for the single-link manipulator. The optimal control scheme was used in the work of **Pal** *et.al.* [12] Trajectory tracking of the end-effector was studied by **Bhat and Miu** [13] & got promising results. Recently, [14]. **Jinjun Shan** [14] studied slewing control of single-link manipulators & developed some control strategies to control the tip-position of the manipulator.

Chen & Yeung [15] used sliding mode control to attenuate the vibrations developed due to the actuation of the motors & used it for the regulation of a one-link flexible robot arm using sliding-mode technique. The sensor based feedback controls were carried out by **Kotnik Yurkovich** *et.al.* [16] for the single-link manipulators. Similar to single-link flexible manipulators, in the 2-link manipulators also, several control schemes were used to attenuate the vibration of the manipulator & take it to the desired objective. Here, while computed torque control was considered by **Bayo** *et.al.* [17], the adaptive control was studied by **Skowronski** in [18]. The former work was continued extended and the optimal control was developed & used in the work of **Lee and Wang** [19].

A PD controller was developed by **Yigit** [20] and sliding control was used in the work of **Yao-Wen Tsai and Van Van Huynh** *et.al.* [21] for the tracking purposes, which also incorporated the output feedback concepts in the sliding mode. A self-tuning concept for a non-rigid arm was used by **Koivo** and **Lee** [22] for the control strategy. **Moallem** *et.al.* [23] carried out an observer based inverse dynamics control strategy which resulted in small tip-position tracking errors (which was a drawback), while maintaining the robust closed-loop performance for a class of multi-link structurally flexible manipulators & they used an integral manifold approach. **Konno** *et.al.* [24] found structural vibration uncontrollable configurations within a 2-link and 3-joint type manipulator's workspace and introduced the modal accessibility index. They also used a singular perturbation technique to control the vibration of the joints.

Hillsley and Yurkovich [25] designed a 2-stage control architecture to achieve accurate end-point position control for point to point movements. **Khorami** *et.al.* [26] developed rigid body based controllers with input pre-shaping for a 2-link flexible manipulators. **Kim** *et.al.* [27] used PZT based smart material to control the vibration of 2-link manipulators. **Yuanchun** *et.al.* [28] developed a robust controller for 2-link manipulators using neural network based quasi-static deflection compensation. **Matsuno and Matsuno** *et.al.* [29] studied the hybrid position and force control of two-DOF flexible manipulators. **Manjunath** worked a lot on the forward kinematic analysis algorithm for articulated machines in his research paper in [30] and used for tacking control.

Control in Robotics : Open problems and future directions towards the research was put forth nicely in the work by **Siciliano** [31]. Some of the concepts in his work were used by us to develop the control strategy. **Åström & Hägglund** [32] worked on some advanced PID Control strategies for different applications & one such applications is being used by us in our work.

Ozgoli [33] worked on the position control for flexible joint robots in presence of actuator saturation in flexible manipulators. **Spong** [34] did novel works on the integral manifold approach to the feedback control of FJRs along with which he produced an excellent survey paper on "The control of FJRs: A Survey w.r.t. the new trends and applications of distributed parameter control systems" in his research survey paper in [35]. Also, he developed some mathematical models for the control of elastic joint robots [36].

Control of FJRs via external linearization approach was proposed by **Lin** *et.al.* in [37], but this involved high computations. **Sweet & Good** [38] re-defined the robot motion control problems w.r.t. the effects of plant dynamics, drive system constraints and user requirements in [39]. **Cesareo and Marino** [40] proposed some novel schemes on the controllability properties of elastic robots, but there were couple of limitations beyond the Hooke's point. A excellent survey paper on the "Control of RLFJRs : A Survey on Back stepping Approach" was put forward by the team of **Dawson** and **Abdallah** in [41].

Brogliato, Ortega, Lozano developed a comparative study on the global tracking controllers for FJMs in [42], but some of the studies were limited only to medium size flexible manipulators. Adaptive Control of FJRs was developed by **Khorasani** in [43]. As **Machida**'s [44] team worked on the precise tele-robotic system for space experiment on ETS-VII, the team led by **Nakamura** *et.al.* [45] worked on the development of a soft manipulator using a smart flexible joint for safe contact with humans and got perfect accuracy to reach the set-point controls. An H_∞ controller synthesis design for the flexible joint robots was developed by **Taghirad** and **Gh. Bakhshi** in [46].

A singular perturbation approach to the control of elastic robots was developed by **Ficola, Marino and Nicosia** in [47], but the perturbation was very complex in nature. Advanced PID Control strategy for a large number of industrial applications was produced by **Strom & Hagglund** in [48]. A state feedback controller for flexible joint robots working in a globally stable approach implemented on DLR's light-weight robots was developed by **Albu-Schäffer** and **Hirzinger** in [49], but couple of applications could not be worked upon with as some of the states were not available for the feedback purposes during the fault conditions, which was one of the drawbacks in their proposed control works.

An, Atkeson and **Hollerbach** [50] developed a model-based control schematic of a typical robotic manipulator, but it could be used only for limited parameters (drawback). A survey of the control of flexible manipulators was done by **Benosman** and **Le Vey** in [51]. This survey paper acted as a ready reckoner for many of the researchers who wanted to pursue research in the field of flexible manipulators. Practical models for practical flexible arms was developed in an experimental approach by **Book** and **Obergfell** in [52], where many of the practical problems like how to build the flexible arm that could be used for experimentation purposes were dealt within. One point to be noted was limitation was there upto few link design only & could not be considered for n-link flexible mechanisms.

Feedforward/feedback laws for the control of flexible robots was developed by **De Luca** *et.al.* in [53], even though it gave good results during the control of position, velocity, acceleration, it was unable to be used to control other hybrid parameters (drawback). The same team worked on the rest-to-rest motion of a one-link flexible arm in [54] and produced novel results, one drawback was their strategy could not be used for multiple link flexible arms. To some aspects the team also worked on the regulation of flexible arms under gravity also as they had inputted the g parameter into the mathematical model in [55]. In fact, De Luca *et.al.* also worked how to obtain a stable inversion control for flexible link manipulators using some matrix inversion strategies.

Robust linear control of flexible joint robot systems for any industrial application was devised by the team of **Elmaraghy, Lahdhiri** and **Ciuca** in [56], even though their strategy gave good results, drawback was - the parameters had to be linearized about an operating point, which made use of some approximations (assumed models). Many people had worked on the linearized models. One amongst them was **Hastings** and **Book** [57] where in their work, they developed linear dynamic model for flexible robotic manipulators, which could be used for any control application. Dynamic learning from adaptive neural control for flexible joint robot with tracking error constraints using high-gain observer was developed by the team of **Zhiguang Chen** et.al. *et.al.* in [58]. Further in another paper, they used the linear quadratic regulator LQR which was using different input parameters for the tuning purposes, which had to be done using the trail & error methods, which was proved to be a lacuna in their work.

MIMO state feedback controller for a flexible joint robot with strong joint coupling was designed by the team of **Tien, Albu-Schaffer & Hirzinger** in [59]. Here, the joint coupling mass was considered which was very much negligible compared to the mass of the links. Dynamic simulation of a leadscrew driven flexible robot arm and controller in the software environment was done by **Chalhoub & Ulsoy** in their research paper in [60]. Recursive Lagrangian dynamics of flexible manipulators was modelled by **Book** [61], but the equations what he had modelled were highly complex in nature.

In any flexible & rigid robots, the 2 types of joints, viz., prismatic & revolute joints could be used. **Muller, Ackermann & Gurgoze** used this concept to modelling and control an elastic robot arm with rotary and prismatic joints in [62], but one drawback was to get a prismatic or a linear joint, the motion conversion had to be used, in the sense, if a rotary motion is given by a motor which produces rotary motion, the linear motion had to be obtained by rack & pinion (conversion from rotary to linear). **Asada & Ma** [63] developed an inverse dynamics model of flexible robot arms, but the drawback was the number of inverse kinematic solutions obtained was more as the number of equations to be solved was greater than the number of unknowns as a result of which multiple solutions was got.

In any robot path planning & trajectory planning operation, the state of the parameters (position, velocity & acceleration) plays a vital role as it is those parameters which has to be tracked & set right. To obtain this, one can use the inverse kinematic models. In this context, the team of **Kwon & Book** [64] developed some inverse dynamic methods yielding flexible manipulator state trajectories for reaching the desired set points.

A number of issues in the dynamics and control of flexible robot manipulators was studied & put forward by **Baruh & Tadikonda** in their work in [65]. The duo listed out a large number of issues in the dynamics and control of flexible robot manipulators in their research article, "Issues in the dynamics and control of flexible robot manipulators (elastic arms)" presented in [65]. Here, they modelled each flexible link independent of the others. The joint displacements were then used as the constraints to develop the equations of motions & further used it for simulations. However, flexibility which has to be considered as a known disturbance was not there in the work (assumed).

Yuan, Book & Huggins developed control models for a multi-link flexible manipulator using a decentralized approach in [66]. **Joaquin Cervera, Alfonso Banos** et.al., worked on the tuning of fractional PID controllers by using quantitative feedback theory (QFT) in [67]. Optimal design of PID controllers using the QFT method was developed by **Zolotas** and **Halikias** in [68]. They used majority of the concepts put in the QFT text book titled, "Quantitative Feedback Theory Fundamentals & Design Theory" by **Houpis & Isaac Horowitz** [69].

A large number of mathematical models & control models in the area of flexible manipulators was developed by **Stig Moberg** in [70]. Many people had used these models for their research work for various applications. Here, he deals with different aspects of modelling and control of flexible, i.e., elastic, manipulators. For an accurate description of a modern industrial manipulator, this work showed that the traditional flexible joint model, described in various literatures, is not sufficient, which was a major drawback. An improved model where the elasticity is described by a number of localized multidimensional spring-damper pairs is therefore proposed by them in order to overcome the lacunas. This model was called as the extended flexible joint model. The main contributions of this work were the design and analysis of identification methods and using the inverse dynamics control methods, for the extended flexible joint model development, further using it for control purposes.

Robust controller design using quantitative feedback theory (QFT) was developed by **Vikas, Chandrakanth & Bhagsen** in [71]. In case of conventional control, if plant parameter changes we cannot assure about the system performance hence it is necessary to design robust control for uncertain plant. The modern control systems such as H_2, H_∞ and μ-synthesis can't handle large uncertainty and are applicable to single input single output (SISO) LTI systems. The drawbacks of conventional and modern control theory are eliminated by classical control theory based method which can handle large parameter uncertainty and works in frequency domain called as QFT.

Robots are now an integral part of automation sector in the modern days, thus indicating the importance of the associated control strategies. In contrast with conventional rigid manipulators, flexible arms offer several benefits in terms of light weight and power efficient structure, safe operation due to reduced inertia, low manufacturing cost & faster movements. In this context a team of authors systematically reviewed the key linear as well as nonlinear techniques to control flexible manipulators in the form of an excellent research paper. The flexibility in link as well as in joint was discussed highlighting the control challenges there. It was anticipated that this in-depth study will be potentially beneficial to all the sections of the researchers doing research in flexible robotics. Going Beyond Rigid Manipulators – A Review of Control of Flexible Robotic Arms was conducted by their mammoth survey / review paper by **Ifrah Maqsood** *et.al.* in [72].

High order sliding mode control of a space robot manipulator was proposed by the team of Arisoy et.al. [73] & showed that higher order sliding mode controller produces better results than the lower ones. **Iqbal Pasha** & their research group worked on the real-time target detection and tracking : A comparative in-depth review of strategies in their research paper in [74] and further used this approach to the use of computer vision inspired real-time autonomous moving target detection, tracking and locking of flexible robots in [75], where the tracking was considered as the main parameter to reach the destination. But, due to inertia, flexibility & due to other factors, 100 % efficiency couldn't be achieved.

Impedance control of flexible robot manipulators was proposed by **Zhao-Hui Jiang** in [76]. The impedance control objective was converted into tracking the end-effector impedance trajectories generated by the designed impedance dynamics. An ideal manifold related to the desired impedance trajectory tracking was designed to prescribe the desirable characteristics of the system. The impedance control scheme was derived to govern the motion of the robot system converging and remaining to the ideal manifold in the presence of parametric uncertainties using Lyapunov theory. But, the concept, when applied to an n-link manipulator was not working due to severe dynamic problems.

Numerous control schemes were carried out by a team of control engineers to control the flexible manipulator for various applications. Some of the hard-core topics taken up were -- Experimental control of flexible robot manipulators, Experimental results on variable structure control for an uncertain robot mode, A mathematical tool for modelling, path planning and control of robot manipulators, Motion control of a robot manipulator in free space based on model predictive control, Improvement of force control in robotic manipulators using sensor fusion techniques, Adaptive neural network based fuzzy sliding mode control of flexible robot manipulator, Simple effective control for robot manipulators with friction, etc..... All these control topics were grouped under a common heading titled,

"Robot Manipulators" by **Ceccarelli** in [77]. In majority of the work considered in the former cases, some of the parameters (such as mass of the sensor & actuator patch attached to plant, sensor noise, etc...) were neglected in the design of the controller.

In [78], **Herman Høifødt** created a dynamic model of a flexible robotic manipulator & did extensive simulation of the same using various methodologies & compared the results of the simulator to the Euler-Lagrange Model & Newton-Euler Formulation with some promising results. Since the derivation of the dynamic models in almost any case involved challenging parameter estimations, the dynamic parameters in their work were estimated quite roughly. A perfect model would contain dynamic effects like the joint friction and joint flexibility, and there would be bounds on the maximum input torque for the motors (not considered). These were the drawbacks in their proposed methods.

Hybrid-adaptive switched control for flexible robotic manipulators interacting with arbitrary surface shapes under multi-sensory guidance was proposed by **Danial Nakhaeinia** in his Ph.D. thesis in [79]. The proposed system utilized both offline and online trajectory planning to achieve fully automated object interaction and surface following with or without contact where no prior knowledge of the objects were available. However, one of the drawback recognized was the RGB-D sensors provided only limited accuracy on the depth measurements and create blind spot when it reached close to surfaces.

Minimum-time trajectory control of a two-link flexible robotic manipulator was researched upon by the team of robotists led by **Schoenwald** *et.al.* in [80], where they analyzed the experimental and simulation results of a minimum-time trajectory control scheme along a specified path in minimum time. Their results indicated that the combination of model-based and error-driven control strategies achieved a closer tracking of the desired trajectory and a better handling of the modelling errors (such as tip payloads) than either strategy alone. LQR & Feed Forward Control was used by them. However, there was some lacunas, such as the feedback of integrated position states to reduce the noise effects as well as formulating the problem in discrete time was not considered. In fact, they could have used an LQG design, whereby a Kalman filter could have been used to estimate the states & alleviate the problem of full state feedback in their proposed LQR method.

Zhijie Liu & Jinkun Liu developed a scheme for the boundary control of a flexible robotic manipulator with output constraints in [81]. Their proposed scheme achieved pre-set performance attributes on position tracking error and the deflection error at the end of the flexible manipulator. The dynamics of the elastic system are represented by PDEs & allowed the errors to converge to an arbitrarily small residual set, with convergence rate larger than a pre-specified value, but one major set-back was the number of computations that took to converge was on the higher side.

In many of the complex applications such as in sheet metal cutting, the cutting angles, paths and forces exerted on the material are also important. Moreover, on surfaces where polishing disks must always be perpendicular to the surface being polished, pre-determined force must be applied. This is possible only by the use of lighter and more flexible robots. **Kilicaslan** *et.al.* used this concept by developing some controllers for the hybrid force and motion control of robots with flexible links & achieved success [82]. They designed and constructed a dynamic 3D force/moment sensor to achieve this & was used to provide a feedback signal of forces and moments exerted by the robotic end-effector using an H_∞ & Adaptive Neuro-Fuzzy controller (ANFIS). However, a comparative study was not performed between the developed sensor and the commercially available ones.

A robust control was designed for flexible joint robot manipulators in [83] by **Jong-guk Yim** *et.al.* which was decomposed into 2 cascaded sub-systems ; a series connection of robot link dynamics and joint dynamics. A recursive design method was used for the controller design. The recursive design procedure was constructive and contained 2 steps. First, a fictitious robust controller for the robot link dynamics was designed as if the link dynamics had an independent control. As the fictitious control, a non-linear H_{inf} control using the energy dissipation is designed in the sense of L_2-gain attenuation from the disturbance caused by uncertainties to performance was considered. Second, a real robust control is designed recursively by using a Lyapunov's second method. The designed robust control was then applied to a 2 DOF robot manipulator with joint flexibilities, but the work done was not compared with others.

Constrained motion control of flexible robot manipulators based on recurrent ANNs was developed by **Tian, Wang & Mao** in [84], where both the contact force exerted by the flexible manipulator and the position of the end-effector contacting with a surface are controlled. Based on the error dynamics of a feedback controller, a learning rule for updating the connection weights of the adaptive RNN model is obtained. Local stability properties of the control system were discussed, but there was no intimation of any unmodelled dynamics in their works, which was a major dis-advantage in their work.

Adaptive neural network output feedback control for flexible multi-link robotic manipulators was touched upon by **Belkacem Rahmani & Mohammed Belkheiri** in [85] in the joint space R^j. Their approach was valid for a class of highly uncertain systems with arbitrary points, but with bounded dimensions. The problem of trajectory tracking was solved through developing a stable inversion for robot dynamics using only joint angles measurement using the concept of Lyapunov's stability criterion & validated for the 2-link flexible arm.

Dynamic modelling and characterisation of a 2-link flexible robot manipulator was developed by **Khairudin, Mohamed & Husain** in [86]. The authors considered a planar 2-link flexible manipulator (R^2) incorporating structural damping, hub inertia and payload that moves in the horizontal plane & developed a dynamic model of the system using a combined Euler-Lagrange and used assumed mode method. Simulation was performed to assess the dynamic model and system responses at the hub and end-point of both links in time and frequency domains. However, the validation of the dynamic model was not done by comparing with an actual 2-link flexible manipulator system, which was a lacuna.

Indirect adaptive fuzzy control for flexible-joint robot manipulators using voltage control strategy was developed by **Fateh Mohammad Mehdi & Souzan Chikashani Mahdi** in [87] for electrically driven flexible-joint robot manipulators in which a novel estimation technique was introduced to estimate some of the uncertainty parameters. Their control strategy was a decentralized control one whereas the torque-based control design was a centralized control one, thus making a hybrid control. The required feedbacks for each controller were the motor current, joint position, joint velocity and motor velocity. But, during the training process, the number of epochs taken was little on the higher side.

Zulfatman, Mohammad Marzuki and Nur Alif Mardiyah [88] proposed a novel 2-link flexible manipulator control scheme using sliding mode control based on the Linear Matrix Inequalities (LMI's). In their work, each model of the 2-link flexible manipulator was successfully formulated from the existing Transfer Function (TF) form into state equations. But, a smaller control gain was producing smaller chattering in the control signal u & for some cases it may be sensitive on parameters variations and disturbances.

Modelling and control of a two-link flexible robot manipulator with unmodelled dynamics was studied by **De Witm Nijmeijer & Mahlouji** in [89]. The flexibility of the structure generates structural vibrations that strongly interferes with rigid-body attitude dynamics. The dynamics of the structure increases drastically when the amount of bodies increase and so far researchers were not capable to develop an adequately dynamical model. Their study described the derivation of a new dynamical model for a 2-link flexible robot manipulator with, usually unconsidered, incorporated effects of practical parameters. Based on this theoretical framework, 3 different controller designs were introduced. To regulate the rigid body motions a proportional-plus-derivative (PD) control method was implemented and a notch filter was introduced to suppress the residual vibrations & one of the major drawback was the settling times taken to settle was on the higher sides.

Control of parallel flexible 5-bar manipulator using QFT was developed by the team of **Sandeep Karande, Nataraj, Gandhi and Manoj Deshpande** in [90]. Control strategies need to be developed to control their vibrational behavior. The work was aimed at analyzing the performance of a QFT based controller in controlling the vibrations in flexible multilink manipulator systems during the time when the actuator was actuated with power, while steering it to a particular set location.

Output feedback control of flexible link manipulator using sliding modes was put forth by **Shailaja Ravindra & Prashant Dixit** in [91]. They presented a Periodic Output Feedback (POF) control for positioning a tip of flexible link manipulator (FLM) using sliding modes. Sliding mode observer (SMO) was designed for estimation of system states. Output estimation error became zero in finite time due to sliding modes. The SMC with SMO was designed for positioning a tip of FLM. The method was validated in simulation as well as in experiment. Also, the same researchers worked on the control of tip position of flexible link manipulator using sliding modes & produced astonishing results.

An Integral Fast Output Sampling (IFOS) control for controlling the tip position of single link Flexible manipulator which is a multivariable system has been presented by the team of **Nikhil Singh & Rajendran** in [92]. Fast Output Sampling (FOS) scheme eliminates the need of an observer required for the estimation of internal states of the system. The gain of the IFOS controller was obtained by solving system of Linear Matrix Inequalities (LMI's) using CVX optimization toolbox. Modelling of Flexible Link Manipulator (FLM) system was also presented. Finally, the IFOS control scheme was compared with an observer based discrete time state feedback strategy & excellent results were obtained.

Ho & Tu developed PID controllers for a flexible-link manipulator in [93]. The authors investigated the application of the H_∞ proportional-integral-derivative (PID) control synthesis method to tip position control of a flexible-link manipulator. To achieve high performance of PID control, this particular control design problem is cast into the H_∞ framework. Based on the recently proposed H_∞ PID control synthesis method, a set of admissible controllers is then obtained to be robust against uncertainty introduced by neglecting the higher-order modes of the link and to achieve the desired time-response.

Adaptive output feedback control of flexible-joint robots using neural networks with the help of a dynamic surface design approach was carried out by the team of Yoo, Park & Choi in their research article in [94]. **Vahid Azimi** worked on the tool position tracking control of a nonlinear uncertain flexible robot manipulator by using robust H_2 / H_∞ controller via T–S fuzzy model in [95], where they used the regional Pole-Placement concept for obtaining the stability & good control. The superiority of the proposed control scheme is finally highlighted in comparison with the Quantitative Feedback Theory (QFT) controller, the QFT controller of order 13, a polynomial controller and the so-called linear Sliding-Mode Controller (SMC) methods.

Malihe Mirshamsi & Mansour Rafeeyan developed QFT control scheme of a 2-link rigid-flexible manipulator in [96], whose first link is rigid and the second is flexible. A piezoelectric patch was attached to the surface of the flexible link for vibration suppression of it. This system is modeled as a non-linear multi-input multi-output (MIMO) control systems whose inputs are two motor torques which are applied on joints and a voltage which is applied on piezoelectric patch, of course good results were obtained.

Amor Jnifene worked on the active vibration control of flexible structures using delayed position feedback in [97], where they uses a simple position control system approach to improve the performance of lightly damped dynamic systems & further used a delayed position feedback signal to actively control the vibrations of flexible structures. A complete analysis of the stability of a single-link flexible manipulator under time delay control was also presented and critical values of time delay for a given controller gain had been determined.

A number of control schemes were developed for the control of flexible robots using different methods by the team of David & Book in their research article in [98]. In [99], the modelling, control & implementation of smart flexible structures was developed by

Manjunath *et.al.* using a number of controller designs, after producing excellent results, the work was further extended to design & develop robust multirate output feedback controllers for smart intelligent flexible structures for a number of cases by **Arunkumar** *et.al.* in [100].

1000's of papers were collected, referred & studied on the chosen research topic, *"Control of n-link flexible robotic manipulators in 3D Euclidean space"* and here only a few of them (base papers) which were being used are being cited & referred to in the references. A comparision of the some (few) of the noteworthy cited researchers/authors were also made regarding the type of control strategy they had used, their advantages & what the drawbacks/lacunas was in their methodology and the entire chronology of items discussed is presented in the table 2.1 below.

Ref. No.	Type of control used	Advantages	Dis-advantages
98	QFT Control	composite control scheme	required loop shaping technique
94	Non-linear output feedback control	compensation the unmatched uncertainties	internal system dynamics more
89	PD control	notch filter was introduced to suppress the residual vibrations	settling times taken to settle was on the higher sides.
88	SM control	smaller chattering in the control signal u (used LMI)	sensitive on parameters variations and disturbances
87	Indirect adaptive fuzzy control	control strategy was a decentralized control one	no. of epochs taken was little on the higher side
84	Recurrent ANN control	good control of tip position	no intimation of any unmodelled dynamics.
83	Robust control	use of recursive design to control 2 modes	not compared with others.
52	Feedback control scheme	build practical models	was limitation was there upto few link design only
58	Joint control	input-estimation approach and LQG method gave good stability factors	was using different turning parameters
63	inverse dynamic control	good flexibility & faster settling times	number of IK solutions obtained was more
57	Tuning control	gave satisfactory steady state response	For each parameter, tuning had to be done.
75	Tracking control	tracking was considered as the main parameter to reach the destination	due to inertia, flexibility 100 % efficiency could not be obtained

76	Impedance control	governed the motion of the robot system effectively, transient response was faster	concept when applied to a *n*-link manipulator was not working due to dynamic problems.
78	Dynamic control	fast response to the tracking inputs due to the use of Euler-Lagrangian method	joint friction and joint flexibility was put as the bounds on the maximum input torque for motors
79	Hybrid-adaptive switched control	utilized both offline and online trajectory planning to achieve fully automated object interaction & surface following with or w/o contact	RGB-D sensors provided only limited accuracy on the depth measurements and create blind spot when it reached close to surfaces.
80	Minimum-time trajectory control	outputs was tracked in a specified path in minimum time to reach set-point	feedback of integrated position states to reduce the noise effects as well as formulating the problem in DT - not considered

Table 2.1 : A comparision of few of the authors works (advantages & lacunas)

2.3 Drawbacks of the existing works

Like this, a large number of researchers had worked on the flexible systems and in fact, only the important works have been presented in this literature survey. In majority of the work done by the different researchers / authors presented in the previous paragraphs, there were lot of disadvantages / burdens / lacunas / drawbacks / deficiencies, for example,

- some thought of just single actuator for the movement of the joint,
- SISO case,
- no noise was considered,
- robustness was not considered,
- many of them used linearized models,
- non-linearized mathematical models was not considered,
- linearization about an operating point was done,
- utilization of traditional methodologies for control purposes,
- rigid links,
- no under-actuation concept,
- multiple sensors of different nature were not tried upon,
- only one type of sensor was considered,
- parallel flexible systems were not considered till date,
- hybrid controllers were not used to track a particular parameter,

and so on & so forth.

2.4 Conclusions

A large number of articles, papers, thesis, reports done by various authors, researchers, engineers, students, faculties were surfed upon studied in brief. Few of the drawbacks [1] – [100] of the works that were carried out by the earlier researchers were considered in our research work, studied in brief & algorithms were developed in order to overcome some of the deficiencies of the existing algos. The research work is verified through effective simulation results in the Matlab-Simulink in order to substantiate the research problem undertaken in comparison with the work done by the earlier authors in the relevant field, in the sense to solve the desired *objective* (question) & arrive at the *outcome* (answer/solution) of the research work.

Chapter – 3

Controller of Flexible Manipulators using PID

In this chapter, the design & development of controllers to control / track few parameters of the flexible manipulators are presented in brief along with the simulation results & finally concludes with the conclusions. To start with, a small insight into the PID controllers is presented.

A Proportional - Integral - Derivative controller (*PID controller or three-term controller*) is a control loop feedback mechanism widely used in industrial control systems and a variety of other applications requiring continuously modulated control in order to control various parameters of a given plant to a set-value.

The PID controller consists of a form of phase lead-lag compensator having one pole at origin and one pole at the infinity. A PID controller can also be split into 2 parts, namely the PI and PD controllers, where the PI controller is a form of phase-lag compensator and the PD controller is a form of phase-lead compensator. A standard PID controller is also known as a *"three-term"* controller, whose transfer function is given by the Eqn. (3.1) as

$$G(s) = K_p + K_I \left(\frac{1}{s}\right) + K_D s \tag{3.1}$$

Tuning a PID controller appears to be conceptually intuitive, yet can be tough in practice if multiple objectives such as short transient and substantial stability are to be achieved. Initially, the properties of PID frequency are analyzed. Then, according to the results, a PID controller is designed that bridges the relation among the proportional gain (K_p), the integral gain (K_i), and the derivative gain (K_d), with the characteristics of closed loop feedback. This relationship is helpful in adjusting the values of PID parameters according to the response of the closed loop system. The 3 steps required for designing the PID controller for the case considered in our research work are …

1. **Step 1 (P)** : Proportional tuning involves correcting a target proportional to the difference. Thus, the target value is never achieved because as the difference approaches zero, so too does the applied correction.
2. **Step 2 (I)** : Integral tuning attempts to remedy this by effectively cumulating the error result from the 'P' action to increase the correction factor.

3. **Step 3 (D)** : Derivative tuning attempts to minimize this overshoot by slowing the correction factor applied as the target is approached.

All the 3 steps are used in this Chapter. In this context, two works are being carried out in the wake of the control of flexible manipulators, which are

Work – 1 : Section 3.1
- Development of the simulink model for a single link flexible manipulator case.
- PID Controller design for the single link flexible manipulator case.
- Tuning of the controller using Ziegler-Nicholas method for a 1-link case.
- Observation of the results for a 1-link case.

Work – 2 : Section 3.2
- Development of the simulink model for a double link flexible manipulator case.
- PID Controller design for the double link flexible manipulator case.
- Tuning of the controller using Ziegler-Nicholas method for a 2-link case.
- Observation of the results for a 2-link case.

Both the works are discussed in greater detail in the forthcoming sections 3.1 & 3.2 respectively.

3.1 Development of the simulink model for a single link flexible manipulator case with results & discussions

In this section, the design & development of the simulink model & the control scheme to control / track the speed of the motor for a 1-link flexible manipulator case is being considered. Controller is designed in the Matlab-Simulink environment for a single link flexible robotic manipulator case.

The developed simulink model is used to regulate the speed of an DC electric motor which is connected to base of the single link flexible manipulator using one degrees-of-freedom PID control with set-point weighting being set in the model & is as shown in the Fig. 3.1. Note that one motor which is attached to the base of the manipulator corresponds to one DOF. We have used the PID controller inside the blocks & the PID equations are converted into block-sets in the simulink diagram as shown in the subsequent diagrams.

The internal block-diagram of the PID controller is shown in the Fig. 3.2. The basic functions such as the step signals, comparators, PID block set & the control loops, scopes, sinks, connectors, sub-blocks, source blocks, etc… are being used from the simulink library with their numerical values being put in the loop inside them.

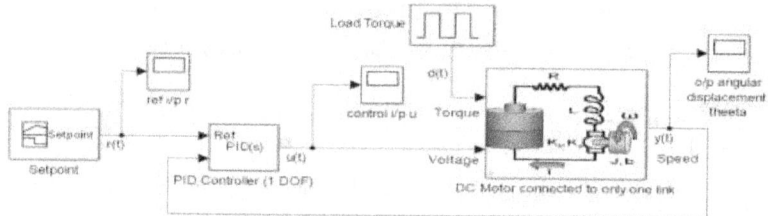

Fig. 3.1 : Simulink model with one DOF PID control of a DC motor which is connected to the base of the flexible single link manipulator

It has to be noted that to observe the different waveforms, scopes are to be connected at relevant places. All these blocks are available in the simulink library, which has to be pulled one by one into the *.mdl file, model to be built and with the correct simulation parameters inside the block, the model has to be run. All these numerical values of the models are incorporated in the PID controller simulink model developed in the Simulink environment for controlling the speed of the DC motor which is connected to the base of the 1-DOF flexible link manipulator.

The single link flexible manipulator is made of high grade aircraft aluminium so that it is light in weight and has good material properties and is of rectangular shape (which can be thought of similar to a wing of an aircraft). In the model shown in the Fig. 1.3, the set-point is set to the single link flexible manipulator for its speed to be regulated/controlled, which is given as the reference input $r(t)$. The PID controller is designed in such a way that the o/p of the PID controller is given as the input to the single link flexible manipulator, i.e., to the DC motor to which the shaft is connected. The K_P, K_i & K_D values are designed & set in such a way so as to tract the given set-point and are inserted into the PID block.

The controller gain constants designed using the compensator formula is given by

$$P(b \times r - y) + I\frac{1}{s}(r - y) + D\frac{N}{1 + N\frac{1}{s}}(c \times r - y) \qquad (3.2)$$

with their values as

$K_P = 0.2$; $K_I = 1.0$; $K_D = 0.06$; Set point weight $b = 1$; Set point weight $c = 1$
No. of filter coefficients $N = 12$; $P =$ Proportional constant ; $I =$ Integral constant
$D =$ Derivative constant ; $S =$ complex variable

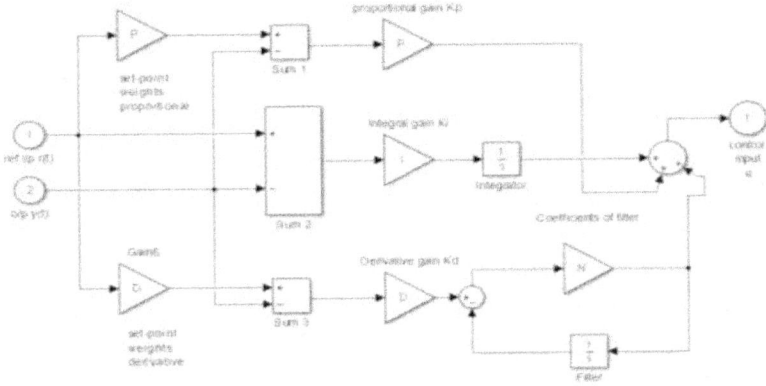

Fig. 3.2 : Internal block diagram of the PID controller for a single link case

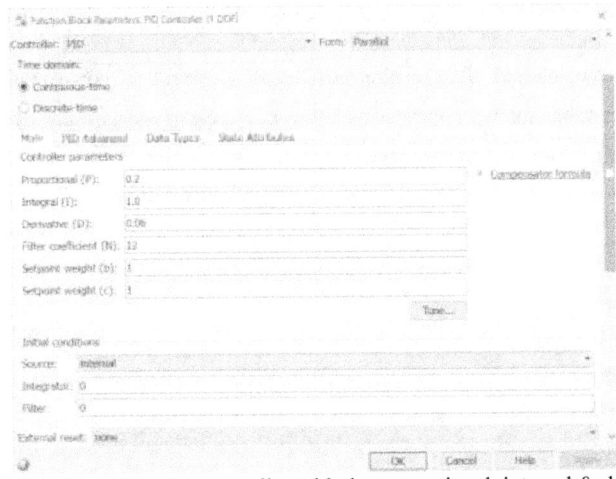

Fig. 3.3 : Tuning of the PID controller with the proportional, integral & derivative constants & PID controller tuning parameters K_P, K_I, K_D values set in the block

All these designed parameters are set in the tuning box parameters shown in the Fig. 3.3, which is obtained by double clicking the PID block. Once, all the parameters are set, next the "*tune*" icon tab is pressed so that the controller is tuned to track the set-points. In contrast to the PID Controller block, the PID Controller (1 DOF) block also provides an extra degree of freedom to allow the researchers to weight the set-point as it passes through the proportional action part and the derivative action part in the controller. The internal block-diagram of the PID Controller (1 DOF) which shows all the 3 individual parts, viz., proportional action, integral action & the derivative action is shown in the Fig. 3.4.

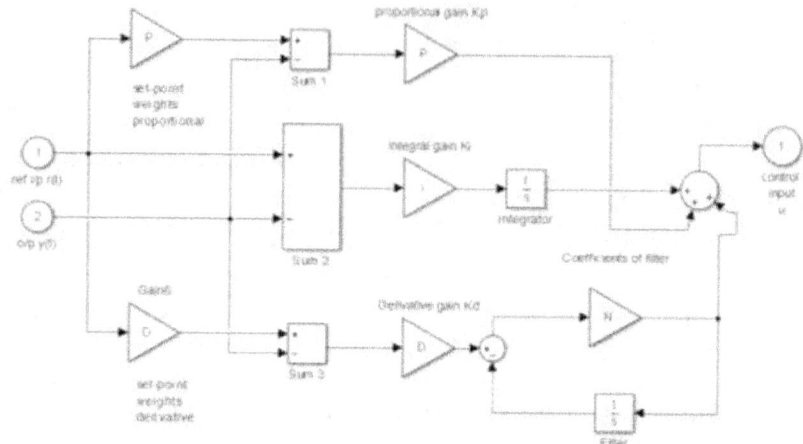

Fig. 3.4 : Internal block diagram of the PID controller for a single link case

From the internal block diagram of the PID controller shown in the Fig. 3.4, the following points can be enunciated as follows. The error signal as seen by the proportional action (from the Fig. 3.4) is given by

$$b \times r - y \tag{3.3}$$

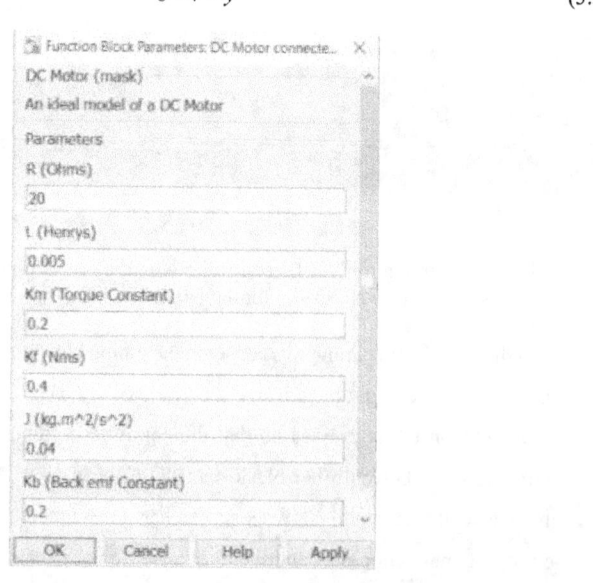

Fig. 3.5 : Setting up of the DC motor parameters in the function block parameter icon

The signal seen by the derivative action (from the Fig. 3.4) is given by

$$c \times r - y \qquad (3.4)$$

The signal seen by the integral action (from the Fig. 3.4) is given by

$$r - y \qquad (3.5)$$

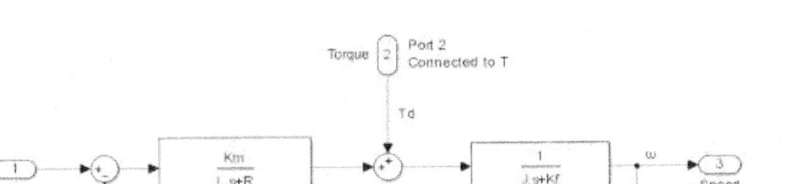

Fig. 3.6 : Internal block-diagram of the DC motor

Generally, the set-point weight in the PID block c is chosen to be 0 in order to prevent undesirable transients that occur during the change in the set-point transitions, which is an effect known as derivative kick. The set-point b affects the overshoot performance of the controller. Generally, a small value of b is taken so that it will reduce the overshoot. However, smaller values of b will also result in slower response due to the changes in the set-point. In our case, we have taken the value of $b = c = 1$ so that the behavioural performance of the PID controller becomes identical to that of a classical PID controller.

Plant (DC motor) : The type of motor used for the displacement is the armature-controlled DC motor. The shaft speed of the motor is controlled by the voltage input to the motor & is diagrammatically shown in the Fig. 3.1. Due to the voltage input, the motor experiences a load torque varying from (0-10 Nm) depending on the voltage input to the DC motor. The specifications of the DC motor is selected as

$R = 20\ \Omega$; $L = 5$ mH ; Motor torque constant $(K_m) = 0.2$;
Field winding constant $(K_f) = 0.4$ NMs ; Moment of Inertia $(J) = 0.04$ Kg.m^2/s^2 ;
Back EMF constant $(K_b) = 0.2$

All these parameters are being set up in the block set which is being pulled down from the simulink library into the developed model. The set parameters are shown in the Fig. 3.5 & the mathematical model is being interpreted into the simulink diagram as shown in the Fig. 3.1.

Observation of the responses, results & discussions :

Once the model is designed, then the simulation has to be carried out for a particular value of time interval. The simulation results shows how the output is being tracked, regulated or controlled with the given set-point. Also, the results shows the efficacy of the methodology developed. The developed PID controller, when put in feedback loop (canonical) with the controller system, showed the following results (presented one after the other subsequently) when the simulink model is run. The responses are observed with the controller. The results itself shows how the tuning is done by selecting the values of K_P, K_I & K_D. In fact, Ziegler-Nicholas method is being used by us for the tuning purposes.

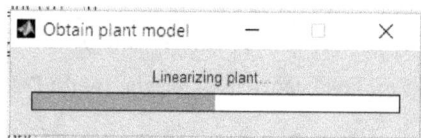

Fig. 3.7 : Result obtained after the tuning of the PID controller

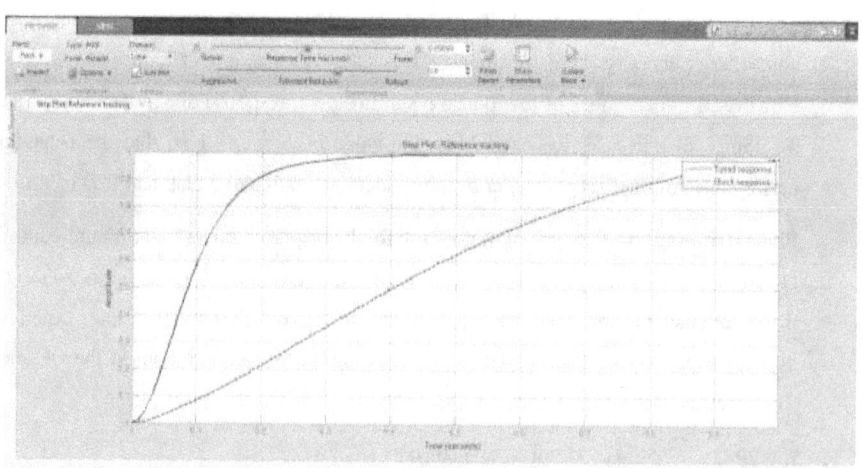

Fig. 3.8 : Plot of step response for the ref. tracking (tuned response & block response)

All these designed parameters are set in the tuning box parameters shown in the Fig. 3.3. Once, all the parameters are set, next the *"tune"* icon tab is pressed so that the controller is tuned to track the set-points. Once, the tune tab is pressed, the plant is linearized & the outputs can be observed in the respective scopes. The plant which is getting linearized is shown in the result in Fig. 3.7 with the plot of reference tracking, set-point tracking in the Fig. 3.8 (step response). The tuned response & the block response are shown in the Fig.

3.8. The quantitative result of the tuned & the tuning parameters is shown in the Fig. 3.9, which is obtained by pressing the show parameter button.

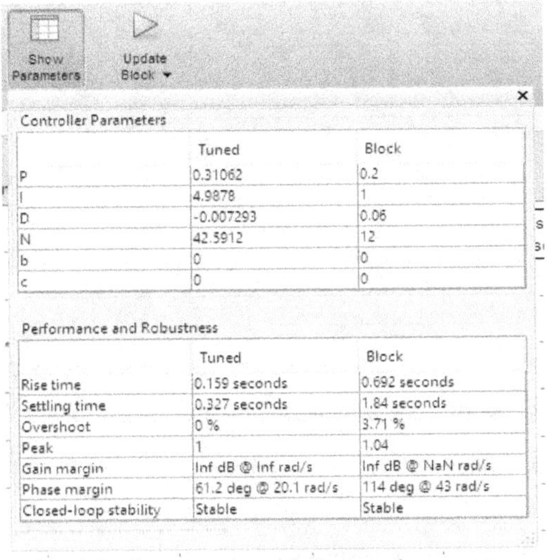

Controller Parameters

	Tuned	Block
P	0.31062	0.2
I	4.9878	1
D	-0.007293	0.06
N	42.5912	12
b	0	0
c	0	0

Performance and Robustness

	Tuned	Block
Rise time	0.159 seconds	0.692 seconds
Settling time	0.327 seconds	1.84 seconds
Overshoot	0 %	3.71 %
Peak	1	1.04
Gain margin	Inf dB @ Inf rad/s	Inf dB @ NaN rad/s
Phase margin	61.2 deg @ 20.1 rad/s	114 deg @ 43 rad/s
Closed-loop stability	Stable	Stable

Fig. 3.9 : Set parameters & the tuned parameters

Plot of the output y, control input u, reference set point r are shown in the simulation results in the Figs. 3.8 to 3.14 respectively. From the simulation results, it can be observed that the output is tracking the set-points which are set. Of course, there is some jerk, perturbations seen at the starting points where there is transition in the set-points. This is due to the non-linearities presented in the system such as dead zone, hysteresis, backlash, friction, resistance, damping, etc.... Each graph has time (in seconds) along the x-axis and y-axis is the amplitude of the controlled output, i.e., the regulated speed. Simulations were done for a particular time period, say 50 seconds.

The set-point signal r, control signal u, and closed-loop response y of the model are shown in Fig. 3.10 – 3.14 respectively. The o/p angular displacement of the motor (speed, ω rads/sec) is shown in the Figs. 3.10 & 3.11 respectively. The torque input to the DC motor to which the link is connected is shown in the Fig. 3.12. The set-point given as input to the PID controller to control the speed (angular displacement) of the DC motor is shown in the Fig. 3.13. The control input u given as input to the PID controller is shown in the Fig. 3.10.

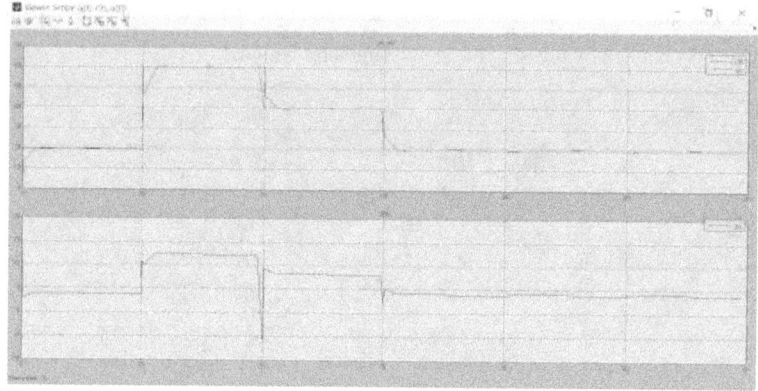

Fig. 3.10 : Output y angular displacement tracking the set-point (ref. i/p, r).....top simulation result along with the control input u .... Bottom simulation result

From the simulation result of the control input u v/s time, it can be clearly that spikes in the control signal u are present during the transition period, which are caused by the aggressive proportional and derivative response to the set-point changes. It can be seen that by modifying the values of b & c weights, the response can be made less aggressive, which has to be obtained by trail & error method by the way of tuning the parameters properly & the fine-tuned response is shown in the Fig. 3.14. The PID Controller (1 DOF) block in Simulink supports the PID control scheme. This block can be used for tracking complex set-point profiles and moderating the impact of sudden set-point changes on control signal transients.

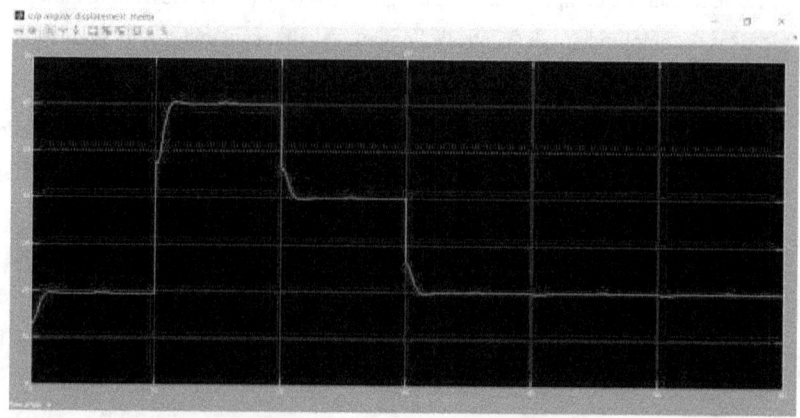

Fig. 3.11 : Output angular displacement of the motor (speed, ω rads/sec)

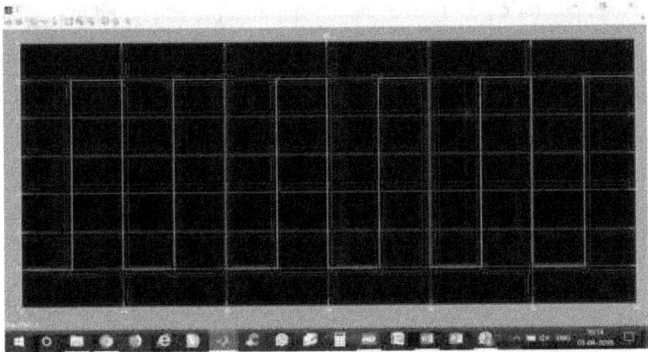

Fig. 3.12 : Torque input to the DC motor to which the link is connected

Fig. 3.13 : Set-point given as input to the PID controller to control the speed (angular displacement) of the DC motor

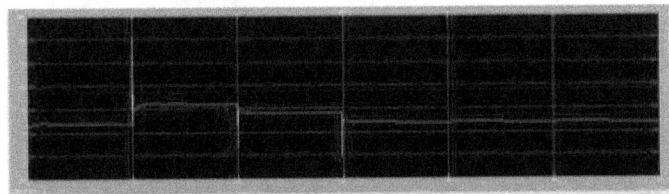

Fig. 3.14 : Control input u given as input to the PID controller

Finally, to conclude, controller was designed for a single link flexible manipulator for the tracking control of the base motor using the set-point & proper tuning of the P, I & D values so that the tracking is done with least error, which can be seen from the output simulated results.

3.2 Development of the simulink model for a two link flexible manipulator case with results & discussions

In this section, the design & development of the simulink model & the control scheme to control / track the speed of the motor for a 2-link flexible manipulator case is being considered. Controller is designed in the Matlab-Simulink environment for a two link flexible manipulator case. The developed simulink model is used to regulate the speed of an DC electric motor which is connected to base of the two link flexible manipulator using 2 degrees-of-freedom PID control with set-point weighting being set in the model & is as shown in the Fig. 3.15. Internal block diagram of the PID controller for a two link case (for the base motor), similar one will be there for a shoulder motor is shown in the Fig. 3.16. Note that one motor at the base corresponds to one DOF, while the other motor kept at the shoulder, i.e., at the end of link 1 serves as another DOF. We have used the PID controller inside the blocks & the PID equations are converted into block-sets in the simulink diagram.

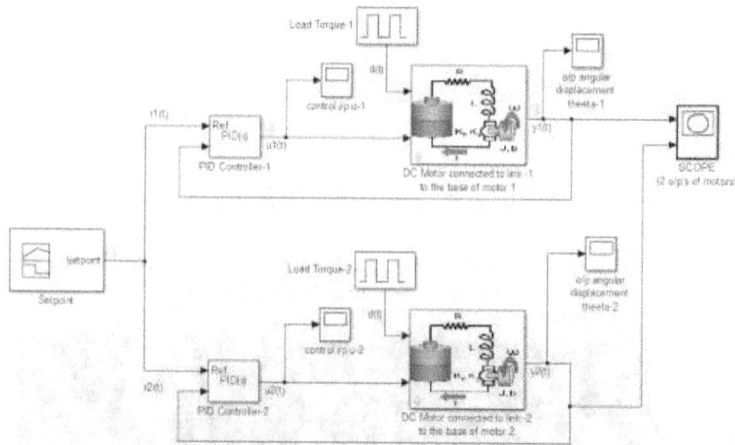

Fig. 3.15 : Developed simulink model of the tracking control of the 2-link flexible robotic manipulator using a single set-point

The internal block-diagram of the PID controller is shown in the Fig. 3.17 for the PID block & the Fig. 3.16 for the motor block for a double link control case separately. The basic functions such as the step signals, comparators, PID block set & the control loops, scopes, sinks, connectors, sub-blocks, source blocks, etc… are being used from the simulink library with their numerical values being put in the loop inside them. It has to be noted that to observe the different waveforms, scopes are to be connected at relevant places. All these blocks are available in the simulink library, which has to be pulled one by one into the *.mdl file, model to be built and with the correct simulation parameters inside the block, the model has to be run. All the numerical values of the models such as the PID constants, motor values, etc…are incorporated in the PID controller simulink model & the motor model block, which is developed in the Simulink environment for controlling the speed of the DC motor/s which are connected to the base of the 2-DOF flexible link manipulator & to the shoulder joint respectively. The length of the link 2 should be < that of the length of the link 1 (similar to the human hand, i.e., the aspect ratio should be < 1).

The dual link flexible manipulator is made of high grade aircraft aluminium so that it is light in weight and has good material properties and is of rectangular shape (*which can be thought of similar to a wing of an aircraft* **or** *the rotating blade of a turbine* **or** *a wind mill*). 2 links are used for a 2 link manipulator along with 2 DC motors. The link-1 is attached to the base of the motor (named as base motor) & the link-2 is attached to the shoulder, i.e., to the end of the link-1 (named as shoulder motor). The manipulator has got 2 links & 2 joints, named as the base & the shoulder joints. The tip of the manipulator is used for doing any manipulation, i.e., an end-effector could be attached to it for doing any manipulation. It has to be noted that the 2-link manipulator is a planar one as the 2 links are parallel to the plane of the work-surface. In the model shown, the set-point is set to the 2-link flexible manipulator for its speed to be regulated/controlled. The K_P, K_i & K_D values are designed & set in such a way so as to tract the given set-point and are inserted into the PID block. It is to be noted that if the actuators, i.e., the motors are controlled properly by developing a sophisticated control scheme, then tracking, positioning, etc... all can achieved with minimal errors.

The controller gain constants designed using the compensator formula is given by

$$P(b \times r - y) + I\frac{1}{s}(r - y) + D\frac{N}{1 + N\frac{1}{s}}(c \times r - y) \qquad (3.6)$$

with their values as

$K_p = 0.3$; $K_I = 1.3$; $K_D = 0.09$; Set point weight $b = 1$; Set point weight $c = 1$
No. of filter coefficients $N = 15$ for the base motor (joint 1 shown in Fig. 1.3)

&

$K_p = 0.25$; $K_I = 1.1$; $K_D = 0.08$; Set point weight $b = 1$; Set point weight $c = 1$

No. of filter coefficients $N = 10$ for the shoulder motor (joint 2 shown in Fig. 1.3)

All these designed parameters are set in the tuning box parameters by double clicking the PID block sets. Once, all the parameters are set, next the "*tune*" icon tab is pressed so that the controller is tuned to track the set-points.

Plant (DC motor) Base motor : The type of motor/s used for the angular displacement is the armature-controlled DC motor. The shaft speed of the motor is controlled by the voltage input to the 2 motor/s individually. Due to the voltage input/s, the motor/s experiences a load torque varying from (0-10 Nm) depending on the voltage input to the DC motor/s. The specifications of the shoulder motor should be < that of the base motor as the shoulder motor has to provide energy only to the link-2 & to the tip, whereas the base motor should be having more than the shoulder ones as it has to bear the entire weight from the base till the tip. The specifications of the DC base motor connected to the joint-1 is selected as

$R = 20\ \Omega$; $L = 5$ mH ; Motor torque constant $(K_m) = 0.2$;
Field winding constant $(K_f) = 0.4$ NMs ; Moment of Inertia $(J) = 0.04$ Kg.m^2/s^2
Back EMF constant $(K_b) = 0.2$.

The specifications of the DC shoulder motor connected to the joint-2 is selected as

$R = 15\ \Omega$; $L = 3$ mH ; Motor torque constant $(K_m) = 0.15$;
Field winding constant $(K_f) = 0.3$ NMs ; Moment of Inertia $(J) = 0.02$ Kg.m^2/s^2 ;
Back EMF constant $(K_b) = 0.16$

All these parameters are being set up in the block set which is being pulled down from the simulink library into the developed model. The set parameters are shown in the Figs. 3.3, 3.7 & 3.9 respectively & the mathematical model is being interpreted into the simulink diagram shown in the Fig. 3.15. Once the model is designed, then the simulation has to be carried out for a particular value of time interval. Result obtained after the tuning of the PID controller is shown in the Fig. 3.18, showing that the tuning is in process.

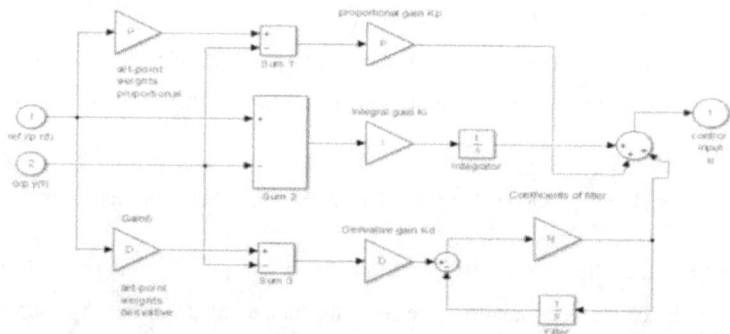

Fig. 3.16 : Internal block diagram of the PID controller for a two link case (for the base motor), similar one will be there for a shoulder motor

The simulation results shows how the output is being tracked, regulated or controlled with the given set-point. Also, the results shows the efficacy of the methodology developed. The developed parallel PID controller for a 2-DOF case, when put in feedback loop (canonical) with the controller system, showed the following results (presented one after the other subsequently) when the simulink model is run. The responses are observed with the controller.

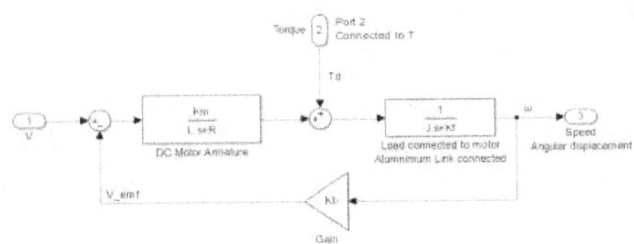

Fig. 3.17 : Internal block-diagram of the DC motor

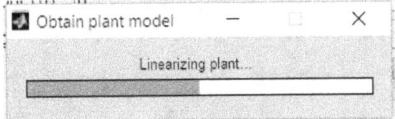

Fig. 3.18 : Result obtained after the tuning of the PID controller

All these designed parameters are set in the tuning box parameters shown in the Figs. 3.3, 3.5 & 3.9 respectively. Once, all the parameters are set, next the "*tune*" icon tab is pressed so that the controller is tuned to track the set-points. Once, the tune tab is pressed, the plant is linearized & the outputs can be observed in the respective scopes. The plant which will then get linearized as shown in the Fig. 3.18. Plot of the output y, control input u, reference set point r are shown in the simulation results in the Figs. 3.19 to 3.20 respectively.

From the simulation results, it can be observed that the output is tracking the set-points which are set. Of course, there is some jerk, perturbations seen at the starting points where there is transition in the set-points. This is due to the non-linearities presented in the system such as dead zone, hysteresis, backlash, friction, resistance, damping, etc.... Each graph has time (in seconds) along the x-axis and y-axis is the amplitude of the controlled output, i.e., the regulated speed. Simulations were done for a particular time period, say 50 seconds.

The set-point signal r, control signal u, and closed-loop response y of the model are shown in Fig. 3.19 for the base motor, whereas the result for the same parameters for the shoulder motor is shown in the Fig. 3.20. From the result of the control input u, we can see that there are some spikes present (sharp rises), which is due to the sudden transitions in the change of voltage levels (from high to low & vice-versa). It can be seen from the output results of the 2 motors stats that for the base motor, more control effort is needed to control it than the shoulder motor as it has to carry the weight of the link-1 + weight of the shoulder motor + weight of link-2 & the payload.

For the shoulder motor, it has to take care of not only controlling the weight of link-2 & the payload, but also it has to take care of the tip. Hence less control effort u needed to control it is less, which can be seen from the output result as shown in the Fig. 3.20. It has to be noted that the control effort required at the starting point for the shoulder motor is just half that of the effort needed at the base motor. The set-point tracking for both the motors will be the same as the main agenda is to control / tract the set point, but the only difference is in the control effort u.

Motor type (joint)	$t = 0\ s$	$t = 10\ s$	$t = 20\ s$	$t = 30\ s$
Base motor	150 mV	−30 mV	130 mV	−38 mV
Shoulder motor	78 mV	35 mV	24 mV	10 mV

Table 3.1 : Quantitative statistics of the control effort needed to control the motor

Note that in our case, both the motors are tracked / controlled separately one after the other. A single PID controller also could be designed in such a way that it can be used to control both the motors (similar to an under-actuated system).

From the simulation result of the control input u v/s time for the 2 motors, it can be clearly that spikes in the control signal/s are present during the transition period, which are caused by aggressive proportional and derivative response to the set-point changes. It can be seen that by modifying the values of b & c weights, the response can be made less aggressive, which has to be obtained by trail & error method by the way of tuning the parameters properly. The PID Controller (2 DOF) block in Simulink supports the PID control scheme. This block can be used for tracking complex set-point profiles and moderating the impact of sudden set-point changes on control signal transients.

Fig. 3.19 : Output simulation results of the tracking control, control input & the set-point reference input for the base motor – 1 to which link 1 is connected
(more control effort needed in this case)

Fig. 3.20 : Output simulation results of the tracking control, control input & the set-point reference input for the base motor – 2 to which link 2 is connected
(less control effort needed in this case)

3.3 Conclusions

Research was carried out on the development of tracking control algorithms for the control of flexible robotic manipulators in the 3 dimensional Euclidean space using PID controllers. A single link manipulator was considered followed by a 2 link manipulator for the simulation purposes as the plant model. Mathematical model of the PID was used to develop the controller to track the speed of the motors in the flexible link manipulators. Software tool '*Matlab + Simulink*' was going to be used to solve the identified problem by developing simulink models & running the same to observe the simulation results and arrive at the expected output results (*goals*). The simulation results show the efficacy of the developed PID controller to control the tracking of the actuators for the single link & dual link manipulators. The 2 works on single link & two link manipulators was also published in 2 reputed IEEE conferences.

Chapter – 4

Design of Controllers using Periodic Output Feedback – POF for 1 & 2-link FMs

In this 2nd contributory chapter, a brief *introduction* about the controller design using the periodic output feedback theory is dealt with aftermath being used to control the various parameters of the 1-link & 2-link flexible manipulator. Once, the controller is designed using the POF concepts (*a type of multi-rate output feedback controller*), the developed controller is put in loop with the plant (1-link or 2-link flexible manipulator) and the control strategy developed is tested for its effectiveness. Matlab tool is being used to develop the control algorithm and observe the results. The research work done is compared with some of the works of the yesteryears. Simulink model for the POF designed controller is also developed & the results are observed. The chapter concludes with the discussion of the simulation results followed by the overall conclusions of the chapter #4.

4.1 A brief insight into the development of the Periodic Output Feedback (POF) Control Design for FMs

Control of flexible manipulators finds a lot of applications in the modern day world, especially in the field of avionics, robotics & the smart intelligent flexible systems. The need for a sophisticated control arises from the output feedback concepts, which has gained a lot of advantages over the traditional control schemes for controlling the various parameters of the plants. One such output feedback algorithm is the periodic output feedback abbreviated as POF. The output of the plant, i.e., the displacement, velocity, acceleration (each one obtained by the differentiation of the earlier ones) is considered as the parameter for control purposes, in our case, the displacement is being considered as the control variable to be tracked upon with. A brief review of which follows in the following paragraphs.

This special control algorithm is developed to cater for the flexibility in the robotic arms, and these algorithms are not that complex, considering that the system is non-linear as generally all flexible systems even though they are non-linear will be linearized about an operating point.

In the work considered, POF is used to regulate the motion of the flexible manipulators & take to a present location and also to track the set-points, curb down the vibrations with less errors. To design the controller, a state space model is obtained first, which is obtained from the first principles starting from the M, B, K parameters. 2 separate cases of the flexible robotic manipulator are presented here with, viz., the SISO (1-link) control, second-the MIMO (2-link) control, in the sense the former has one actuator – so, 1 input & 1 output, whereas the latter has two actuators – so, 2 inputs & 2 outputs. Open loop (OL) & closed loop (CL) responses are also observed. Simulations are carried out with & without the controller to show the authenticity of the proposed control strategies in comparision with the work done by other researchers [91] & [97].

The POF control algorithm actually needs the plant system to be completely controllable & be observable, thus giving the desired performances even in the presence of un-measured disturbances. The performance of the periodic output feedback controller will be tested on the 1 & 2 link flexible manipulator & could be generalized for a n-link flexible manipulator case also, which could be treated as future works. The POF control design proposes a methodical procedure for a robust controller where the control signal u is going to be constrained to be a linear function of most of the recent output measurements of the plant (plant output), but the output is generally allowed to vary at discrete instants of time between the measurements. In this POF controller, the input r at a particular moment of time will depend on the plant output y at a time prior to that moment, mainly at the start of sampling time τ.

POF concepts can be applied to design the controller in which the input u is changed several times in one output sampling interval τ. The problem of control design using piecewise constant periodic output feedback is considered here, because complete pole assignment is not possible using static output feedback. Such a control algorithm can stabilize a much larger class of systems than static output feedback. Since the feedback gains are piecewise constant, this method can be easily implemented to get desired results. The method will not need any assumption with the exception of a general assumption that the system should be completely controllable and observable.

Chammas and Leondes showed in the form of an algorithm / control law as "*if a system is controllable and observable, then for almost all output sampling rates, any self-conjugate pole configuration can be assigned to the discrete time closed loop system by piecewise constant output feedback, provided the number of gain changes is not less than the system's controllability index*". A control design could be proposed such that the stabilizing output injection gain matrix G is obtained for any given system under consideration. The problem could finally lead to the development of a POF gain that realizes this designed output injection gain matrix. Anyhow, the resulting controller even though is stabilizing, tends to give poor closed loop behavior and inter-sample oscillations. Hence, to avoid this, an optimization has to be done with some additional constraints by solving the closed loop parameters & to get an optimized error free outputs.

The control problem of pole assignment by piecewise constant output feedback was studied in brief for the first time by the group of Werner, Chammas and Leondes for LTI systems considering all the infrequent observations. There, they told that if a system is completely controllable and observable, then almost for all the output sampling rates, it can be shown that by using a periodically time-varying piecewise constant output feedback gain, the poles of the DT control system can be completely assigned arbitrarily (*within the natural restriction that they should be located symmetrically w.r.t. the negative real axis in the s-plane*), provided the number of gain changes during one output sampling interval is not < the controllability index γ of the flexible system. Since the feedback gains are piecewise constants, their method could easily be implemented, guarantees the closed loop stability and indicated a new possibility in controller design. Such a control law can stabilize a much larger class of systems than by the use of static output feedback technique.

Now, to start with, we consider a linear time invariant (LTI) continuous time (CT) system modelled by the Eqn. (4.1) as

$$\begin{aligned}\dot{x}(t) &= Ax(t) + Bu(t) \\ y(t) &= Cx(t) + Du(t)\end{aligned} \quad (4.1)$$

where A, B, C, D, x, u, y are the system matrix, input matrix, output matrix, transmission matrix, state variable, input variable & the output variable of the plant (1-link & 2-link flexible system) & are constant matrices of appropriate dimensions.

Now, we discretize the above CT system sampled at Δ interval and the sampled system is given by $(\Phi_\tau, \Gamma_\tau, C)$ and is named as the Δ system. Once again, we discretize the above CT system sampled at τ interval and the sampled system is given by (Φ, Γ, C) and is named as the τ (tau) system. We once again assume that the τ (tau) system is completely observable & the Δ (delta) system is completely controllable. Now, the output y can be measured at different time instants of $t = k\tau$, where $k = 0, 1, 2, 3, \ldots$.

Next step is to consider the constant hold functions (piecewise) since they are more suitable for implementation purposes and to get the error free output. An output-sampling interval τ has to be divided into N sub-intervals of width $\Delta = \dfrac{\tau}{N}$. It has to be noted in this context that the hold function is to be held constant during these sub-interval. The value of N should be $\geq \gamma$, which is the controllability index of the discretized system. After all these modifications done, then, the control law will become

$$u(t) = K_l y(k\tau), \ldots\ldots\ldots (k\tau + l\Delta) \leq t \leq (k\tau + (l+1)\Delta), \quad K_{l+N} = K_l \qquad (4.2)$$

for $l = 0, 1, 2, \ldots, (N-1)$. It should be noted that a continuous sequence of N POF gain matrices given by $\{K_0, K_1, K_2, \ldots\ldots K_{N-1}\}$ when it is entered into the control law equation will generate a time varying piece-wise constant periodic output feedback gain given by $\mathbf{K}(t)$ for $0 \leq t \leq \tau$. Now, in order to obtain the POF gain sequence $\mathbf{K}$, let us consider the delta system as given by the Eqn. (4.3) as

$$\begin{aligned} x(k+1) &= \Phi x(k) + \Gamma u(k) \\ y(k) &= C x(k) \end{aligned} \qquad (4.3)$$

Now, let us consider the POF gain matrix as given by the Eqn. (4.4) as

$$\mathbf{K} = \{K_0, K_1, K_2, \ldots\ldots K_{N-1}\}^T \qquad (4.4)$$

Next step is to calculate the control effort $\mathbf{u}$ to be applied to the system, i.e., $u(t)$ using the control law equation. Finally, the control effort after lot of substitutions and compilations, the control $\mathbf{u}(k\tau)$ will become

$$\mathbf{u}(k\tau) = \mathbf{K} y(k\tau) = \begin{bmatrix} u(k\tau) \\ u(k\tau + \Delta) \\ \vdots \\ u(k\tau + \tau + \Delta) \end{bmatrix} \qquad (4.5)$$

After simplifications, the tau (τ) system could be formulated as given in Eqn. (4.6) as

$$\begin{aligned} x(k\tau + \tau) &= \Phi^N x(k\tau) + \Gamma \mathbf{u}(k\tau) \\ y(k\tau) &= C x(k\tau) \end{aligned} \qquad (4.6)$$

where $\Gamma = \left[\Phi^{N-1}\Gamma,\ \Phi^{N-2}\Gamma,\ \Phi^{N-3}\Gamma, \ldots, \Gamma \right]$.

The next step is to apply the control law given by $u(t) = K_t y(k\tau)$, i.e., $\mathbf{K} = y(k\tau)$ is substituted for $\mathbf{u}(k\tau)$, the closed loop system of the plant is given by Eqn. (4.7) as

$$x(k\tau + \tau) = \left(\Phi^N + \Gamma \mathbf{K} C \right)(k\tau) \qquad (4.7)$$

Finally, this control problem has taken the form of a static output feedback problem. This equation suggests that an output injection gain G can be found out such that

$$\rho\left(\Phi^N + GC \right) < 1 \qquad (4.8)$$

where $\rho(.)$ will be denoting the spectral radius. By observability, one can always choose an output injection gain matrix G so as to achieve the self-conjugate set of Eigen Values of the CLCS (closed loop control system)'s matrix given by $\left(\Phi^N + GC \right)$ and from $N \geq \upsilon$. Then, if one solves the equation $\Gamma \mathbf{K} = G$, then we can find the periodic output feedback gain $\mathbf{K}$, once this gain $\mathbf{K}$ is given as input to the controller, the control action is initiated & the system reaches the stability point in no time, thus gaining stability and reducing the vibrations.

The problem with controllers obtained in this way is that, although they are stabilizing and achieve the desired closed loop behavior at the output sampling instants, they may cause an excessive oscillation between inter-sampling instants. The controller obtained using the above equation may give the desired behaviour, but might require excessive control action and the POF gains obtained may be very high. To reduce this effect, we relax the condition that $\mathbf{K}$ exactly satisfy the linear equation of $\Gamma \mathbf{K} = G$ and include a constraint on it. Solving $\Gamma \mathbf{K} = G$ may give a POF gain that is higher in magnitude, amplifying the noise in a system. Hence, the conditions on $\mathbf{K}$ is imposed during the controller design. These restrictions on $\mathbf{K}$ are posed as LMI problem. Thus, we arrive at the following inequalities (LMI's) as

$$\begin{aligned} \|\mathbf{K}\| &< \rho_1 \\ \|\Gamma \mathbf{K} - G\| &< \rho_2 \end{aligned} \qquad (4.9)$$

Here, in the above equations, small value of ρ_2 corresponds to a good approximation of the POF controller gain $\mathbf{K}$ in the sense that stability is not impaired and small value of ρ_1 corresponds to having a low value of noise sensitivity. Further, the compliment of

Schur's criterion can be employed to the above linear equations by using Linear Matrix Inequalities often called as the LMIs for the optimization of the gains so that good gain will not saturate the system as given by the LMI model as shown in the Eqn. (4.10) as

$$\begin{bmatrix} \rho_1^2 I & \mathbf{K} \\ \mathbf{K}^T & -I \end{bmatrix} < 0$$

$$\begin{bmatrix} \rho_2^2 I & (\Gamma \mathbf{K} - G) \\ (\Gamma \mathbf{K} - G)^T & -I \end{bmatrix} < 0 \quad (4.10)$$

By using the LMI toolbox of Matlab, the optimized value of the gain $\mathbf{K}$ can be found out. It should be noted that in this context, the optimized value of this POF gain K requires only piecewise constant and will definitely be easier to implement in the real time applications.

4.2 Control design for Single link flexible manipulator

It should be noted that in this case, a single flexible link is connected to the hub where a single actuator is used for actuation purpose. Refer Figs. 1.1-1.3 for the diagrammatic representation of the flexible system along with its specifications in the Table 1.1. The actuator is actuated upon by a control effort u such that the desired set-point is reached. In the sense, when the single link gets actuated, the motor as well as the flexible link starts vibrating. At the tip of the flexible link, one displacement sensor is placed, which is used to give the feedback to the system and could be used for tracking purpose. This displacement x (θ) could be considered as one of the state variable, which is used for tracking and bringing back to the desired position in no time. In other words, there is one actuator (motor)-input & one sensor (used to obtain the displacement of the end-effector-output as such the 1-link flexible system can be considered as a SISO system.

The first task in designing POF controller is to fix the sampling time (τ). An external force v (voltage is applied to the actuator) for a particular duration of t sec at the hub of the flexible link & once the motor is actuated, the system starts to move & hence is subjected to vibrations and the open loop impulse response (plot of tip outputs y as a function of t) of the plant is observed. The maximum bandwidth for all the tip displacement sensor & motor-actuator locations (at the base) on the flexible system are calculated (here, the second mode of the overall flexible plant) and then by using the existing empirical rules for selecting the sampling interval based on bandwidth, approximately 10 times of the maximum second mode of the flexible plant is selected, i.e., τ is selected as τ = 0.004 s & the number of sub-intervals N is chosen to be 10.

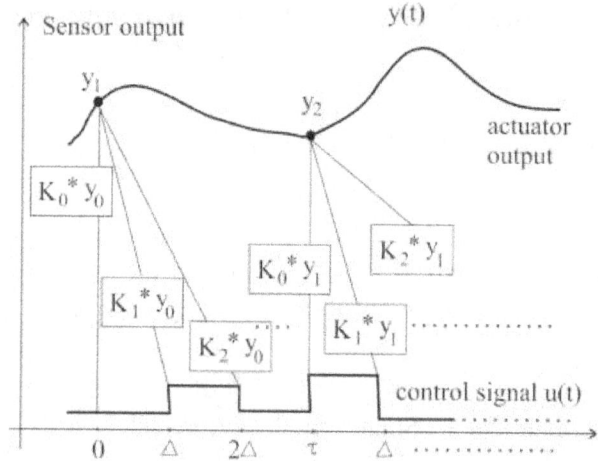

Fig. 4.1 : Graphical illustration of the POF control law developed

The POF control law is graphically displayed in the Fig. 4.1. Next, the given CT system is discretized & the τ & Δ systems are obtained. It is observed that both the τ & Δ systems are controllable and observable. In our case, the controllability index is given by 4 & the observability index is also obtained as 4, in the sense that all the states are controllable & observable. Next, the stabilizing output injection gains G are obtained such that the Eigen values of the τ - DT system is inside the unit circle and is stable.

Next the Δ - DT system is considered and the POF gain **K** is found out be solving the control law equation after taking the sensor output, followed by the LMI optimization concepts such that we get an optimal value of the POF gain so that it reduces the amplitude of the control effort u before taking the end of the flexible link to the desired position. The closed loop response, i.e., the displacement $y(\theta)$ is observed. It could be observed that the closed loop response settles at a faster rate compared to the open loop response.

The designed periodic output feedback controller is put in the loop with the simulated plant and the closed loop step response, i.e. response and variation of the control effort 'u' with time 't' is observed. The output injection gain matrix G is obtained by using the pole placement technique (place command in Matlab). G is obtained such that the Eigen values of the closed loop system $\left(\Phi^N + GC \right)$ lies inside the unit circle and the response of the system has a good settling time. It should be noted that the effect of the state feedback is realized by a multirate output feedback gain **K**. The Open Loop Injection Gain (OLIG), G is obtained as

$$G = \begin{bmatrix} -0.3456 \\ 0.2344 \\ -0.5679 \\ 0.3458 \end{bmatrix} \quad (4.11)$$

The obtained output injection gain G is further used to design the POF controller for the original system. Then, the POF gain **K** can be determined by solving the equation (gamma K = G). But, the magnitude of this POF gain obtained **K** is very large and hence Actually, the gains obtained by the exact solutions are very large and they may cause large variation of inputs during an output sampling instants, which may also require more control effort u exceeding the hardware limitations and cause more SNR to occur.

$$\mathbf{K} = \begin{bmatrix} -10.2346 \\ 7.3324 \\ -9.4567 \\ 8.4582 \end{bmatrix} \quad (4.12)$$

Moreover with large feedback gains, the flexible system becomes more sensitive to noise. So, an LMI optimization procedure is used to get the optimized value of the POF gain **K**. The value of the performance matrices in the LMI concepts, i.e., Q, R, and P are tuned properly to get the optimized POF gain vector **K**, where the gains are having less values so that input saturation will not occur. This gain vector **K** is having lesser magnitude & further it is found that the closed loop response of the original system with POF controller gain **K** gives a stable and desired optimal behavior. These POF gains **K** are of much smaller magnitude than the POF gains obtained before the optimization procedure. Obviously the control effort u is very much reduced and the variation of input during an output sampling interval is also less, thus not taxing the system, further involving less computations. The POF controller designed by the above method requires only constant gains and its implementation is very much easier.

$$\mathbf{K} = \begin{bmatrix} -0.2346 \\ 0.3324 \\ -0.4567 \\ 0.4582 \end{bmatrix} \quad (4.13)$$

The value of P, Q & R chosen (tuned) to obtain the optimal response in the performance index calculation of $J(k)$ was 1.0, 0.05 & 0.008 respectively. Codes are developed in the Matlab environment as .m files. The developed .m files are run, the test input is given as the input to the developed Matlab code & after running the simulation, the simulation results are observed for the contribution #2 and are presented as shown in the Figs. 4.2 – 4.3 respectively for the single link flexible manipulator case. Also, the bode plot in the frequency domain with & without the control effort is also shown here in this context for the sake of authentication to show the control effect. It has to be noted in this case that the gains are of dimensions (1×4) as it is a SISO case (single input single output as there is only one motor & one output displacement) & the state space model (A) is of (4×4).

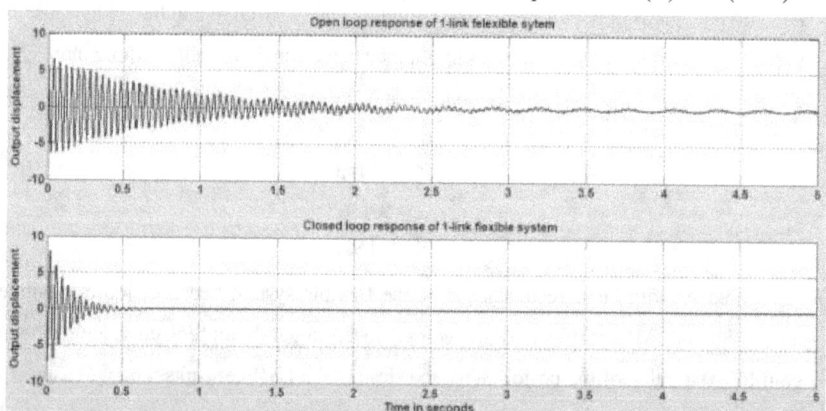

Fig. 4.2 : Open loop & closed loop response of a 1-link flexible manipulator

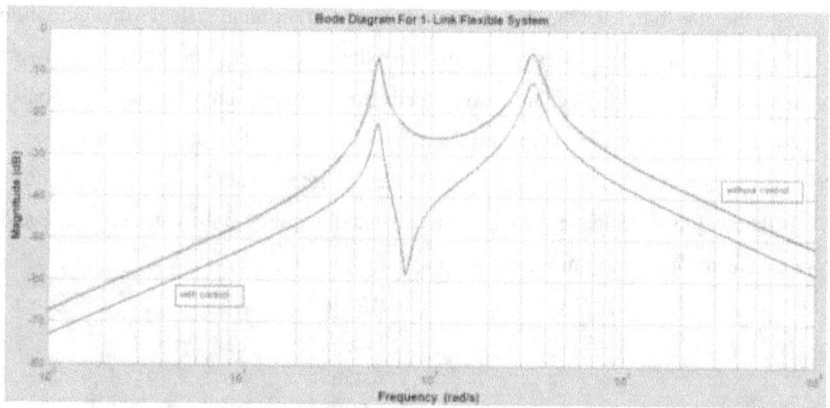

Fig. 4.3 : Bode diagram of a 1-link flexible manipulator to show effectiveness before & after the POF control

4.3 Control design for dual link flexible manipulator

It should be noted that in this case, a dual / double / 2-flexible links are connected to obtain the overall plant structure so that the 1st motor or the 1st actuator is connected to the hub, which is termed as the base motor or base actuator to which the link 1 is connected and to the end of the link-1, another 2nd actuator or the 2nd motor is connected, which is termed as the link-2. Refer Fig. 1.1-1.3 for the diagrammatic representation of the flexible system with its specifications in Table 1.1. Our ROI (region of interest) is the tip of the link 2, which is to be controlled. The actuator/s are actuated upon by a control effort u_1 & u_2 such that the desired set-points are reached.

In the sense, when both the link gets actuated, the motor/s as well as the flexible links starts vibrating. At the tip of the flexible link-2, one displacement sensor is placed, which is used to give the feedback to the system and could be used for tracking purpose. Also, at the tip of the link-1 another displacement sensor-1 is attached for feedback purposes. This displacement $x_1(\theta)$ of the link 1 & $x_2(\theta)$ of the link 2 could be considered as the state variables, which is used for tracking and bringing back to the desired position in no time as a result of which the entire system becomes a 2 input, 2 output system or a MIMO system. In other words, there are two actuators, viz., (motor 1)-base motor & (motor 2)-shoulder motor.

The first task in designing POF controller is to fix the sampling time (τ). An external force v (voltage is applied to the actuator) for a particular duration of t sec at the hub of the flexible link & once the motor is actuated, the system starts to move & hence is subjected to vibrations and the open loop impulse response (plot of tip outputs y as a function of t) of the plant is observed. The maximum bandwidth for all the sensor & motor/actuator locations on the flexible robotic manipulator system are calculated (here, the second mode of the flexible plant) and then by using the existing empirical rules for selecting the sampling interval based on bandwidth, approximately 10 times of the maximum second mode of the plant is selected, i.e., τ is selected as $\tau = 0.004$ s & the number of sub-intervals N is chosen to be 10.

Next, the given CT system is discretized & the τ & Δ systems are obtained. It is observed that both the τ & Δ systems are controllable and observable. Next, the stabilizing output injection gains G are obtained such that the Eigen values of the τ - DT system is inside the unit circle and is stable. Next the Δ - DT system is considered and the POF gain K is found out be solving the control law equation after taking the sensor output, followed by the LMI optimization concepts such that we get an optimal value of the POF gain so that it reduces the amplitude of the control effort u before taking the end of the flexible link to the desired position. The closed loop response, i.e., the displacement y (θ) is observed for the two links, i.e., y_1 & y_2 respectively. It could be observed that the closed loop response settles at a faster rate compared to the open loop response.

The designed periodic output feedback controller is put in the loop with the simulated plant and the closed loop step response, i.e., response and variation of the control effort 'u' with time 't' is observed. The output injection gain matrix G is obtained by using the pole placement technique (place command in Matlab). G is obtained such that the Eigen values of the closed loop system $\left(\Phi^N + GC\right)$ lies inside the unit circle and the response of the system has a good settling time. It should be noted that the effect of the state feedback is realized by a multirate output feedback gain $\mathbf{K}$. The OIG, G is obtained as

$$G = [G_1 \quad G_2] = \begin{bmatrix} -0.9345 & 0.2334 \\ 0.5678 & -0.4523 \\ -0.3334 & 0.3944 \\ 0.9982 & 0.4562 \end{bmatrix} \quad (4.14)$$

The obtained output injection gain G is further used to design the POF controller for the original system. Then, the POF gain $\mathbf{K}$ can be determined by solving the equation (gamma * K = G). But, the magnitude of this POF gain obtained $\mathbf{K}$ is very large and hence Actually, the gains obtained by the exact solutions are very large and they may cause large variation of inputs during an output sampling instants, which may also require more control effort u exceeding the hardware limitations and cause more SNR to occur.

$$\mathbf{K} = [\mathbf{K}_1 \quad \mathbf{K}_2] = \begin{bmatrix} -10.2239 & -9.8876 \\ 8.7654 & -7.3241 \\ -6.3452 & 5.3452 \\ 4.2345 & -3.2213 \end{bmatrix} \quad (4.15)$$

Moreover with large feedback gains, the flexible system becomes more sensitive to noise. So, an LMI optimization procedure is used to get the optimized value of the POF gain $\mathbf{K}$. The value of the performance matrices in the LMI concepts, i.e., Q, R, and P are tuned properly to get the optimized POF gain vector $\mathbf{K}$, where the gains are having less values so that input saturation will not occur. This gain vector $\mathbf{K}$ is having lesser magnitude & further it is found that the closed loop response of the original system with POF controller gain $\mathbf{K}$ gives a stable and desired optimal behavior. These POF gains $\mathbf{K}$ are of much smaller magnitude than the POF gains obtained before the optimization procedure. Obviously, the control efforts, i.e., u_1 & u_2 is very much reduced and the variation of input during an output sampling interval is also less, thus not taxing the system, further involving less computations. The POF controller designed by the above method requires only constant gains and its implementation is very much easier.

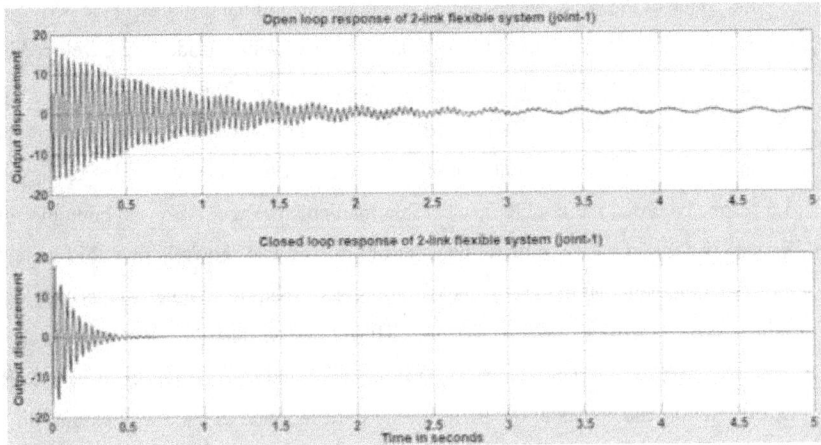

Fig. 4.4 : Open loop & closed loop response of a 2-link flexible manipulator for joint 1 at the base joint

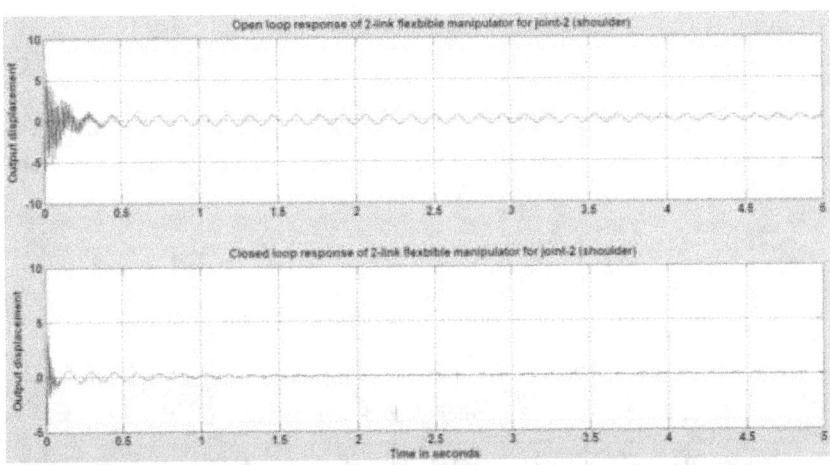

Fig. 4.5 : Open loop & closed loop response of a 2-link flexible manipulator for joint 2 at the shoulder joint

$$\mathbf{K} = \begin{bmatrix} \mathbf{K}_1 & \mathbf{K}_2 \end{bmatrix} = \begin{bmatrix} -0.2345 & 0.3988 \\ 0.3982 & -0.2333 \\ -0.4123 & 0.1118 \\ 0.1123 & -0.4112 \end{bmatrix} \quad (4.16)$$

The value of P, Q & R chosen (tuned) to obtain the optimal response in the performance index calculation of $J(k)$ was 2.0, 0.1 & 0.016 respectively. Codes are developed in the Matlab environment as .m files. The developed .m files are run, the test input is given as the input to the developed Matlab code & after running the simulation, the simulation results are observed for the contribution #2 and are presented as shown in the Figs. 4.4 – 4.6 respectively for the double link flexible manipulator case. Also, the bode plot in the frequency domain with & without the control effort is also shown here in this context for the sake of authentication to show the control effect. It has to be noted in this case that the gains are of dimensions (2 × 4) as it is a MIMO case 2 input 2 output as there are two motors & two output displacement). Also, it was found that the values of P, Q & R are exactly double the values of the performance indices found in the 1-link case & the state space model (A) is of (4 × 4) dimension.

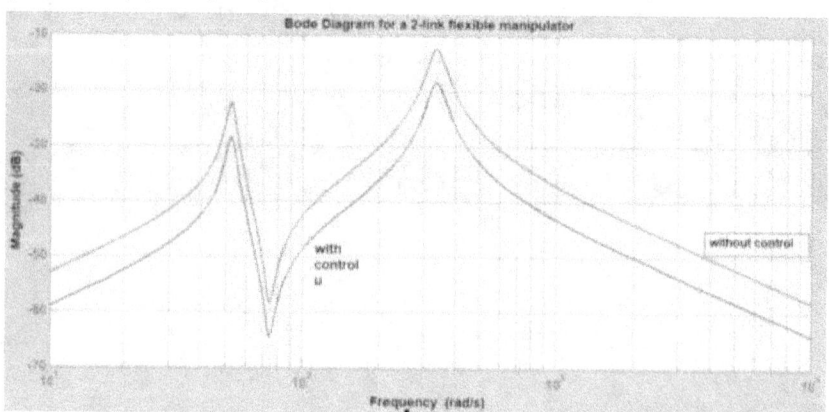

Fig. 4.6 : Bode diagram of a 2-link flexible manipulator to show effectiveness before & after the POF control

4.4 Development of the POF simulink model

The POF controller is also developed in the simulink environment as shown in the Fig. 4.7. The simulink model is constructed using sub-systems, sources, scopes, sinks, comparators, gain blocks, sample and hold circuits, multiplier blocks & the connectors. All these mentioned blocks are available in the simulink modelling library. Apart from these, various toolboxes such as control system tool box, optimization tool box, signal processing tool boxes available in the simulink library is being used. Various parameters are to be set in the different blocks that are used in the development of the simulink model. Simulation time taken is 0 to 5 seconds (varying-any time can be taken). For the discretization purposes, the time, i.e., sample time used in the simulation is 1 ms. The developed simulink model is the same for both the types of flexible systems & only the number of sub-systems will be different. The model is run for the given requisite simulation time & the same results obtained as shown in the Matlab Simulation outputs is observed & hence the results are not shown here for the sake of convenience.

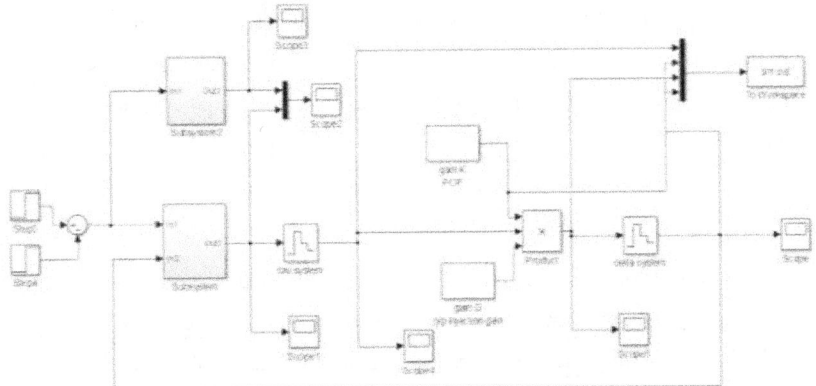

Fig. 4.7 : Simulink model for design of a POF controller for a 1-link & 2-link flexible robotic manipulator system

4.5 Conclusions

Research was conducted on the set-point control, end-point displacements, vibration control of the joints of the tip-1 & tip-2 of the single link and double link flexible manipulator. State space model of the 1 & 2 link system was developed. The CT state space model was discretized @ tau & delta sampling intervals and the discretized systems was obtained for developing the controller. Periodic Output Feedback controllers were designed for the flexible manipulators to control the joint 1 of – link 1 of 1-link manipulator & the joints 1 & 2 of – the 2 link flexible manipulator. The various responses are obtained for each of the state space models of the 2 individual systems.

Through the simulation results, it is inferred that when the plant is placed with this designed controller, the plant performs well, the vibrations die out quickly, the output reaches the set-point quickly in the form of the closed loop response performing much better than the open loop response. From the responses of the single link & 2-link flexible manipulator, it is observed that the output displacement of the link-1 is less than compared to that of the 2-link case as it has to bear drive the link-1 & 2 *plus* the motor-2 weight also. In the 1-link case, the effort is ±20 units, whereas in the 2-link case, the effort is ±15 units for the same base motor.

It could be also noted that in the 2-link case, the effort required for the base motor is ±10 units as it has to drive a less load (only the link-2) *plus* the end-effort mass of the payload. It can also be noted that control effort u required from the controller gets reduced if the actuator is moved away from the fixed end to the end-effector point. A small magnitude of the control signal is sufficient to control the joint-2 of the 2-link case. An output injection gain for each discrete model of the 2 systems is obtained such that poles are placed inside the unit circle at appropriate locations and the CL system has very good settling time.

A very small magnitude of control input u is required to control the actuator as it is moved away from the base compared to the 1-link and 2-link case as a result of which less effort has to be put by the controller. The magnitude of the impulse response (closed loop) of both the continuous and the discrete time system is less compared to 1-link & 2-link case. Also, the close loop response characteristics with G and **K** are the best & settle out quickly. Thus, the observations are made with and without the controller to show the control effect.

From the simulations, it was observed that without control the transient response was unsatisfactory, takes more time to settle and with control, the closed loop response is. Unlike static output feedback, the POF control technique always guarantees the stability of the closed loop system which can be observed from the simulation results. The designed POF controller requires constant gains and hence may be easier to implement in real time application. The controllability & observability index obtained is 4 so that all the states are controllable & observable.

One excellent advantage of the methodology #2 developed is that the computation time required for processing & getting the output is just within 3-4 seconds, which shows the advantage of our proposed method over the others [91] & [97]. One advantage of the 2-link mechanisms is that it can cover a greater area of the work-space & the tip or the end-effector can be tracked in a bigger dimension, i.e., R^2 area as the flexible manipulator works in the planar environment ($x - y$ place).

The work is carried out for both the step & sinusoidal excitations & here only the Matlab & Simulink's simulation results for step excitation is shown for the sake of convenience.

Chapter – 5

Design of Controllers using
Fast Output Sampling Feedback – FOS for 1 & 2-link FMs

In this 3rd contributory chapter, a brief *introduction* about the controller design using the fast output sampling feedback theory is dealt with aftermath being used to control the various parameters of the 1-link & 2-link flexible manipulator. Once, the controller is designed using the FOS concepts (*a type of multi-rate output feedback controller, i.e., sampling @ 2 different rates, viz., τ & Δ*), the developed controller is put in loop with the plant (1-link or 2-link flexible manipulator) and the control strategy developed is tested for its effectiveness. Matlab tool is being used to develop the developed control algorithm and to observe the results. The research work done is compared with some of the works of the yesteryears [91] & [97]. Simulink model for the FOS designed controller is also developed & the results are observed for both the cases. The chapter concludes with the discussion of the simulation results followed by the overall conclusions of the chapter no. #5.

5.1 A brief insight into the development of the Fast Output Sampling Feedback (FOS) Control Design for FMs

Control of flexible manipulators finds a lot of applications in the modern day world, especially in the field of avionics, robotics & the smart intelligent flexible systems. The need for a sophisticated control arises from the output feedback concepts, which has gained a lot of advantages over the traditional control schemes for controlling the various parameters of the plants. One such output feedback algorithm is the fast output sampling feedback abbreviated as FOS. The output of the plant, i.e., the '*displacement*', velocity, acceleration (each one obtained by the differentiation of the earlier ones) is considered as the parameter for control purposes, in our case, the displacement is being considered as the control variable to be tracked upon with. A brief review of which follows in the following paragraphs.

This special multirate control algorithm is developed to cater for the flexibility in the robotic arms, and these algorithms are not that complex, considering that the system is non-linear as generally all flexible systems even though they are non-linear will be linearized about an operating point.

In the work considered, FOS is used to regulate the motion of the flexible manipulators & take to a present location and also to track the set-points, curb down the vibrations with less errors. To design the controller, a state space model is obtained first, which is obtained from the first principles starting from the **M**, **B**, **K** parameters. 2 separate cases of the flexible robotic manipulator are presented here with, viz., the SISO (1-link) control, second-the MIMO (2-link) control, in the sense the former has one actuator – so, 1 input & 1 output, whereas the latter has two actuators – so, 2 inputs & 2 outputs. Open loop (OL) & closed loop (CL) responses are also observed with SFB gain F & with the FOS gain **L**. Simulations are carried out with & without the FOS controller to show the authenticity of the proposed control strategies in comparision with the work done by other researchers [91] & [97].

The FOS control algorithm actually needs the plant system to be completely controllable & be observable, thus giving the desired performances even in the presence of un-measured disturbances. The performance of the fast output sampling feedback controller will be tested on the 1 & 2 link flexible manipulator & could be generalized for a n-link flexible manipulator case also, which could be treated as *future works*.

The FOS control design proposes a methodical procedure for a robust controller where the control signal u is going to be constrained to be a linear function of most of the recent output measurements of the plant (plant output), but the output is generally allowed to vary at discrete instants of time between the measurements. In this FOS controller, the input u at a particular moment of time will depend on the plant output y at a time prior to that moment, mainly at the start of sampling time Δ. In the case of a SISO system, one control variable u_1 is sufficient, where as in a MIMO case, 2 control variables u_1 & u_2 are required.

FOS concepts can be applied to design the controller in which the input u is changed several times in one output sampling interval Δ. The problem of control design using piecewise constant fast output sampling feedback is considered here, because complete pole assignment is not possible using static output feedback. Such a control algorithm can stabilize a much larger class of systems than static output feedback. Since the feedback gains are piecewise constant, this method can be easily implemented to get desired results. The method will not need any assumption with the exception of a general assumption that the system should be completely controllable and observable (*all states should be in a position to be controlled & all the states should be seen, i.e., output measurements can be made*).

A standard result in control theory says that the poles of a LTI controllable system can be arbitrarily assigned by state feedback. In many cases, the entire state vector is not directly available for feedback purposes. Hence, it is desirable to go for an output feedback design. The static output feedback problem is one of the most investigated problems in the control theory and applications.

One reason why the static output feedback has received so much attention is that it represents the simplest closed loop control that can be realized in practical situations. However, no results are available till today which show that complete pole assignment is possible using static output feedback. Practicing state feedback and optimal output feedback controllers needs careful consideration in smart structure application area like the space structures, because the state feedback controller needs the availability of the entire state vector or need estimator.

In the state feedback case, the optimal control law requires the design of a state observer. This increases the implementation cost and reduces the reliability of the control system. Another disadvantage of the observer based control system is that even slight variations of the model parameters from their nominal values may result into significant degradation of the closed-loop performance.

The static output feedback requires only the measurement of the system output, but there is no guarantee of the stability of the closed loop control system. Although the stability of the closed loop system can be guaranteed using the state feedback, the same is not true using static output feedback. So, if a system, for example, smart cantilever beam, in this case, has to be stabilized using only the output feedback (states may not be available for measurement purposes), one can resort to fast output sampling feedback, which is static in nature as well, guarantees the closed loop stability. Here, the value of the input at a particular moment depends on the output value at a time prior to this moment (namely at the beginning of the period).

The problem of fast output sampling was studied by Werner and Furuta for LTI systems with infrequent observations. They have shown that the poles of the DT control system could be assigned arbitrarily (*within the natural restriction that they should be located symmetrically with respect to the real axis*) using the fast output sampling technique. Since the feedback gains are piecewise constants, their method could easily be implemented and indicated a new possibility. Such a control law can stabilize a much larger class of systems.

The proposed flow-chart for the controller design using the FOS methodology is depicted in the Fig. 5.1.

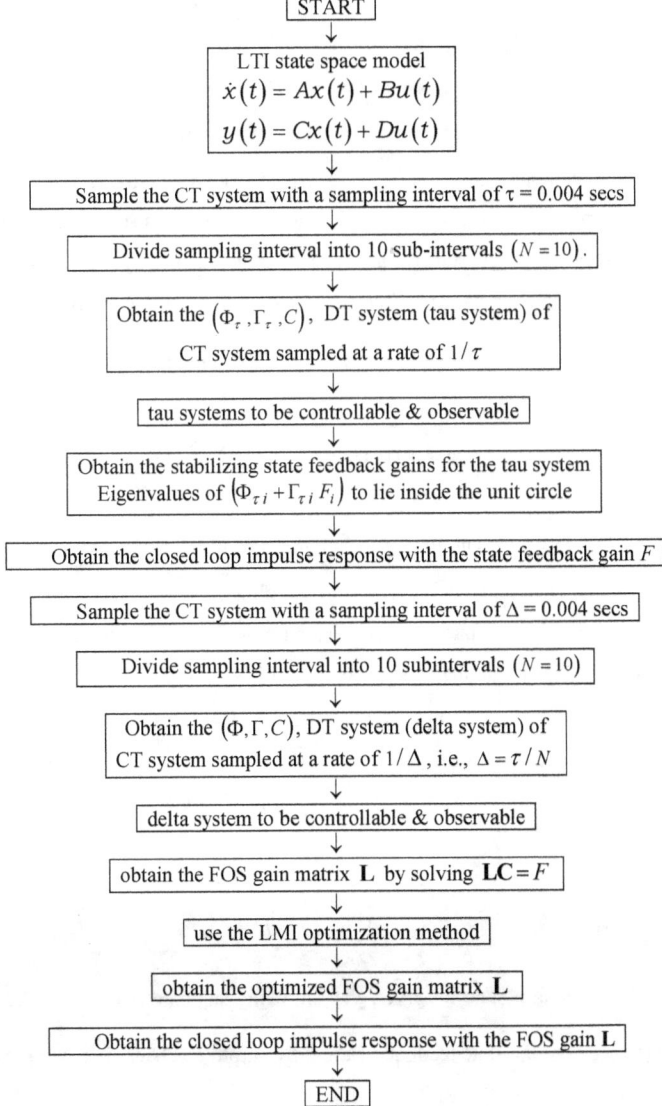

Fig. 5.1 : Proposed flow-chart for the controller design using the FOS methodology

Now, to implement this proposed flow-chart shown in the Fig. 5.1 for the controller design using the FOS concepts on the 1-link & 2-link flexible robotic manipulator, we consider a plant described by a LTI state space model given by the Eqn. (5.1) as

$$\dot{x}(t) = Ax(t) + Bu(t)$$
$$y(t) = Cx(t) + Du(t) \quad (5.1)$$

where $x \in \Re^n$, $x \in \Re^m$, $x \in \Re^p$, $A \in \Re^{n \times n}$, $B \in \Re^{n \times m}$, $C \in \Re^{p \times n}$, A, B, C are constant matrices and it is assumed that (A, B) is controllable and (C, A) is observable. Assume that output measurements are available at time instants $t = k\tau$, where $k = 0, 1, 2, 3, \ldots$ It is to be noted that A, B, C, D, x, u, y are the system matrix, input matrix, output matrix, transmission matrix, state variable, input variable & the output variable of the plant (1-link & 2-link flexible system) & are constant matrices of appropriate dimensions. Now, construct a discrete linear time invariant system from these output measurements at sampling rate $\frac{1}{\tau}$ (sampling interval of τ secs). The system obtained so is called as the τ system and is given by the Eqn. (5.2) as

$$x((k+1)\tau) = \Phi_\tau x(k\tau) + \Gamma_\tau u(k\tau),$$
$$y(k\tau) = C x(k\tau), \quad (5.2)$$

where $\Phi_\tau, \Gamma_\tau, C$ are constant matrices of appropriate dimensions. Assume that the plant is to be controlled by a digital computer, with sampling interval τ and zero order hold (ZOH) and that a sampled data state feedback design has been carried out to find a (SFB) state feedback gain F such that the closed loop control system (CLCS)

$$x(k\tau + \tau) = (\Phi_\tau + \Gamma_\tau F) x(k\tau) \quad (5.3)$$

has desirable properties. Here,

$$\Phi_\tau = e^{A\tau} \quad (5.4)$$

&

$$\Gamma_\tau = \int_0^\tau e^{As} ds \, B. \quad (5.5)$$

Instead of using a state observer, the following sampled data control can be used to realize the effect of the state feedback gain F by output feedback. Let $\Delta = \frac{\tau}{N}$ and consider

$$u(t) = \begin{bmatrix} L_0 & L_1 & \cdots & L_{N-1} \end{bmatrix} \begin{bmatrix} y(k\tau - \tau) \\ y(k\tau - \tau + \Delta) \\ \vdots \\ \vdots \\ y(k\tau - \Delta) \end{bmatrix}, \qquad (5.6)$$

i.e.,

$$u(t) = L \; y_k \qquad (5.7)$$

for $k\tau \le t \le (k+1)\tau$, where the matrix blocks L_j represent the output feedback gains and the notation L, y_k has been introduced for convenience. Note that $\dfrac{1}{\tau}$ is the rate at which the loop is closed, whereas the output samples are taken at the times N - times faster rate $\dfrac{1}{\Delta}$.

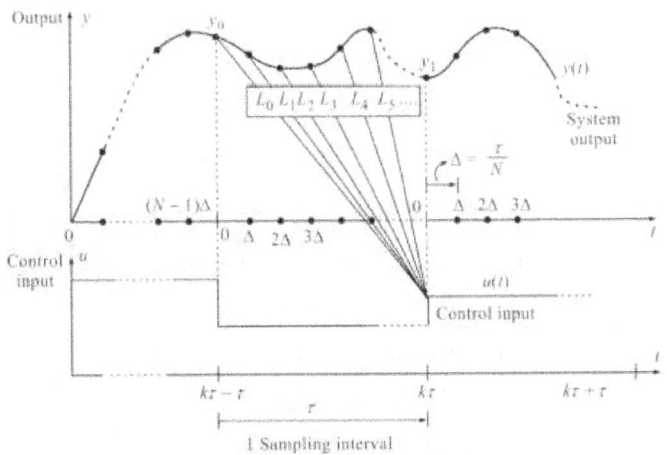

Fig. 5.2 : Graphical illustration of FOS feedback method

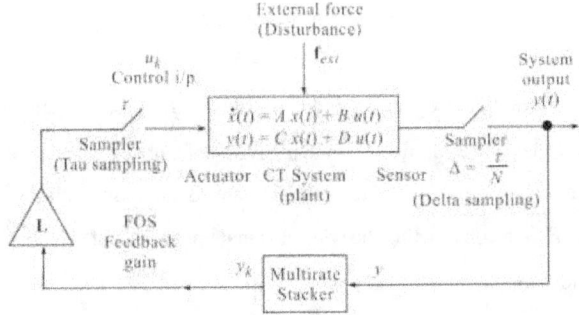

Fig. 5.3 : Block-diagrammatic representation of FOS feedback method of control strategy

This control law is illustrated in the Fig. 5.2 along with the proposed block diagrammatic representation of the FOS control strategy in the Fig. 5.3. To show how a fast output sampling controller in Eqn. (5.7) can be designed to realize the given sampled data state feedback gain for a controllable and observable system (A, B, C), we construct a fictitious, lifted system for which the Eqn. (5.6) can be interpreted as static output feedback. Let (Φ, Γ, C) denote the system in Eqn. (5.1) sampled at the rate $\frac{1}{\Delta}$ and is called as the delta system. Consider the discrete time system having at the time $t = k\tau$, the input $u_k = u(k\tau)$, the state $x_k = x(k\tau)$ and the output y_k as

$$x_{k+1} = \Phi_\tau x_k + \Gamma_\tau u_k,$$
$$y_{k+1} = C_0 x_k + D_0 u_k, \qquad (5.8)$$

where

$$\mathbf{C}_0 = \begin{bmatrix} C \\ C\Phi \\ \vdots \\ C\Phi^{N-1} \end{bmatrix}, \quad \mathbf{D}_0 = \begin{bmatrix} 0 \\ C\Gamma \\ \vdots \\ C\sum_{j=0}^{N-2} \Phi^j \Gamma \end{bmatrix}. \qquad (5.9)$$

Assume that the state feedback gain F has been designed such that $(\Phi_\tau + \Gamma_\tau F)$ has no Eigen values at the origin. Then, assuming that in the interval $k\tau \leq t \leq (k\tau + \tau)$,

$$u(t) = F x(k\tau), \qquad (5.10)$$

one can define the fictitious measurement matrix,

$$\mathbf{C}(F, N) = (\mathbf{C}_0 + \mathbf{D}_0 F)(\Phi_\tau + \Gamma_\tau F)^{-1}, \qquad (5.11)$$

which satisfies the fictitious measurement equation

$$y_k = \mathbf{C} x_k. \qquad (5.12)$$

For **L** to realize the effect of F, it must satisfy the equation

$$\mathbf{LC} = F. \qquad (5.13)$$

Let υ denote the observability index of (Φ, Γ, C). It can be shown that for $N \geq \upsilon$, generically **C** has full column rank, so that any state feedback gain can be realized by a fast output sampling gain **L**. If the initial state is unknown, there will be an error $\Delta u_k = u_k - F x_k$ in constructing the control signal under the state feedback. One can verify that the closed-loop dynamics are governed by

$$\begin{bmatrix} x_{k+1} \\ \Delta u_{k+1} \end{bmatrix} = \begin{bmatrix} \Phi_\tau + \Gamma_\tau F & \Gamma_\tau \\ 0 & \mathbf{LD}_0 - \Gamma_\tau \end{bmatrix} \begin{bmatrix} x_k \\ \Delta u_k \end{bmatrix}. \tag{5.14}$$

To see this, apply the coordinate transformation,

$$T = \begin{bmatrix} I & 0 \\ F & I \end{bmatrix} \tag{5.15}$$

to the Eqn. (5.16)

$$\begin{bmatrix} x_{k+1} \\ u_{k+1} \end{bmatrix} = \begin{bmatrix} \Phi_\tau & \Gamma_\tau \\ \mathbf{LC}_0 & \mathbf{LD}_0 \end{bmatrix} \begin{bmatrix} x_k \\ \Delta u_k \end{bmatrix} \tag{5.16}$$

and use the Eqn. (5.11). Thus, one can say that the Eigen-values of the closed-loop system under a fast output sampling control law given in Eqn. (5.11) are those of $(\Phi_\tau + \Gamma_\tau F)$ together with those of $(\mathbf{LD}_0 - \Gamma_\tau)$. This suggests that the state feedback F should be obtained so as to ensure the stability of both $(\Phi_\tau + \Gamma_\tau F)$ and $(\mathbf{LD}_0 - F\Gamma_\tau)$. The system in Eqn. (5.8) is stable if and only if F stabilizes (Φ_τ, Γ_τ) and the matrix $(\mathbf{LD}_0 - F\Gamma_\tau)$ has all its Eigen values inside the unit circle.

The problem with controllers obtained in this way is that, although they are stabilizing and achieve the desired closed loop behavior in the output sampling instants, they may cause an excessive oscillation between sampling instants. The fast output sampling feedback gains obtained may be very high. To reduce this effect, we relax the condition that L exactly satisfy the linear equation (5.13) and include a constraint on the gain L. Thus, we arrive at the following in Eqns. (5.17)-(5.20).

$$\|L\| < \rho_1, \quad \|\mathbf{LD}_0 - F\Gamma_\tau\| < \rho_2, \quad \|\mathbf{LC} - F\| < \rho_3. \tag{5.17}$$

This can be formulated in the form of Linear Matrix Inequalities (LMI's) as given in the Eqns. (5.17) to (5.20) as

$$\begin{bmatrix} -\rho_1^2 I & \mathbf{L} \\ \mathbf{L}^T & -I \end{bmatrix} < 0, \tag{5.18}$$

$$\begin{bmatrix} -\rho_2^2 I & \mathbf{LD}_0 - F\Gamma_\tau \\ (\mathbf{LD}_0 - F\Gamma_\tau)^T & -I \end{bmatrix} < 0, \tag{5.19}$$

$$\begin{bmatrix} -\rho_3^2 I & \mathbf{LC} - F \\ (\mathbf{LC} - F)^T & -I \end{bmatrix} < 0. \tag{5.20}$$

In this form, the LMI optimization toolbox is used for the synthesis of **L**. Thus, finally by using the LMI toolbox of Matlab, the optimized value of the gain **L** can be found out. It should be noted that in this context, the optimized value of this FOS gain **L** requires only piecewise constants and will definitely be easier to implement in the RT applications.

5.2 Control design for Single link flexible manipulator

It should be noted that in this case, a single flexible link is connected to the hub where a single actuator is used for actuation purpose. Refer Figs. 1.1-1.3 for the diagrammatic representation of the flexible system with its specifications in Table 1.1. The actuator is actuated upon by a control effort u such that the desired set-point is reached. In the sense, when the single link gets actuated, the motor as well as the flexible link starts vibrating. At the tip of the flexible link, one displacement sensor is placed, which is used to give the feedback to the system and could be used for tracking purpose. This displacement x (θ), which is nothing but the output of the system y could be considered as one of the state variable, which is used for tracking and bringing back to the desired position in no time. In other words, there is one actuator (motor)-input & one sensor (used to obtain the displacement of the end-effector-output as such the 1-link flexible system can be considered as a SISO system.

The first task in designing FOS controller is to fix the sampling time (Δ). An external force v (voltage is applied to the actuator) for a particular duration of t sec at the hub of the flexible link & once the motor is actuated, the system starts to move & hence is subjected to vibrations and the open loop impulse response (plot of tip outputs y as a function of t) of the plant is observed. The maximum bandwidth for all the tip displacement sensor & motor-actuator locations (at the base) on the flexible system are calculated (here, the second mode of the overall flexible plant) and then by using the existing empirical rules for selecting the sampling interval based on bandwidth, approximately 10 times of the maximum second mode of the flexible plant is selected, i.e., Δ is selected as $\Delta = 0.004$ s & the number of sub-intervals N is chosen to be 10.

The FOS control law is graphically displayed in the Fig. 5.2 with its block diagrammatic counterpart in the Fig. 5.3. Next, the given CT system is discretized & the Δ & τ systems are obtained. It is observed that both the Δ & τ systems are controllable and observable. In our case, the controllability index is given by 4 & the observability index is also obtained as 4, in the sense that all the states are controllable & observable. Next, the stabilizing FOS feedback gains L are obtained (after obtaining the state feedback gain F) such that the Eigen values of the Δ - DT system is inside the unit circle and is stable.

Next the τ - DT system is considered and the FOS gain **L** is found out be solving the control law equation after taking the displacement output from the sensor place at the end of the end-effector of the FM, followed by the LMI optimization concepts such that we get an optimal value of the FOS gain so that it reduces the amplitude of the control effort u before taking the end of the flexible link to the desired position. The closed loop response, i.e., the displacement $y(\theta)$ is observed. It could be observed that the closed loop response settles at a faster rate compared to the open loop response.

The designed fast output sampling feedback controller is put in the loop with the simulated plant and the closed loop step response, i.e. response and variation of the control effort 'u' with time 't' is observed. The state feedback gain matrix F is obtained by using the pole placement technique (place command in Matlab). F is obtained such that the Eigen values of the closed loop system $(\Phi_\tau + \Gamma_\tau F)$ lies inside the unit circle and the response of the system has a good settling time, i.e., stabilizing state feedback gains are obtained for the tau systems such that the eigenvalues of $(\Phi_\tau + \Gamma_\tau F)$ are inside unit circle. It should be noted that the effect of the state feedback is realized by a multirate output feedback gain **L**. The State Feedback Gain (SFG), F is obtained as

$$F = \begin{bmatrix} -0.2134 & 0.3354 & -0.4659 & 0.5758 \end{bmatrix} \quad (5.21)$$

The closed loop impulse response of the 1-link flexible manipulator system with the state feedback gain F is also observed. The obtained state feedback gain F is further used to design the FOS controller for the original system. Then, the FOS gain **L** can be determined by solving the linear gain equation $LC = F$. But, the magnitude of this FOS gain obtained **L** is very large and hence Actually, the gains obtained by the exact solutions are very large and they may cause large variation of inputs during an output sampling instants, which may also require more control effort u exceeding the hardware limitations and cause more SNR to occur.

$$\mathbf{L} = \begin{bmatrix} -10.7134 & 9.3358 & -10.4699 & -10.4699 \end{bmatrix} \quad (5.22)$$

Moreover with large feedback gains, the flexible system becomes more sensitive to noise. So, an LMI optimization procedure is used to get the optimized value of the FOS gain **L**. The value of the performance matrices in the LMI concepts, i.e., ρ_1, ρ_2 & ρ_3 are tuned properly to get the optimized FOS gain vector **L**, where the gains are having less values so that input saturation will not occur. This gain vector **L** is having lesser magnitude

& further it is found that the closed loop response of the original system with FOS controller gain **L** gives a stable and desired optimal behavior.

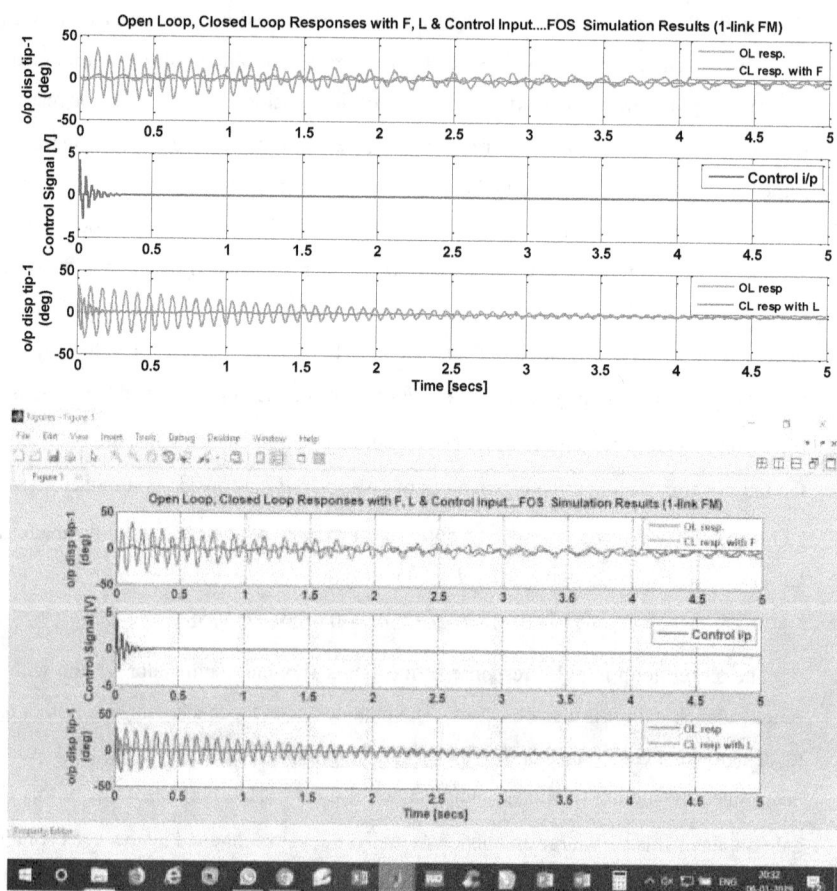

Fig. 5.4 : Open loop & closed loop response of a 1-link flexible manipulator with SFB gain F & FOS gain **L** with control input u

These FOS gains **L** are of much smaller magnitude than the FOS gains obtained before the optimization procedure. Obviously, the control effort u is very much reduced and the variation of input during an output sampling interval is also less, thus not taxing the system, further involving less computations. The FOS controller designed by the above method requires only constant gains and its implementation is very much easier.

$$\mathbf{L} = \begin{bmatrix} -0.0352 & 0.1346 & 0.1346 & 0.6357 \end{bmatrix} \qquad (5.23)$$

The value of ρ_1, ρ_2 & ρ_3 chosen (tuned) to obtain the optimal response in the performance index calculation of *norms* of ρ was $\rho_1 = 0.01$, $\rho_2 = 0.002$ & $\rho_3 = 0.0003$ respectively. Codes are developed in the Matlab environment as .m files. The developed .m files are run, the test input is given as the input to the developed Matlab code & after running the simulation, the simulation results are observed for the contribution #3 and are presented as shown in the Figs. 5.4 & 5.5 respectively for the single link flexible manipulator case. Also, the bode plot in the frequency domain with & without the control effort is also shown here in this context for the sake of authentication to show the control effect. It has to be noted in this case that the gains are of dimensions (1×4) as it is a SISO case (single input single output as there is only one motor & one output displacement) & the state space model (A) is of (4×4). From the frequency plot, it can be observed that the dB value is reduced with control.

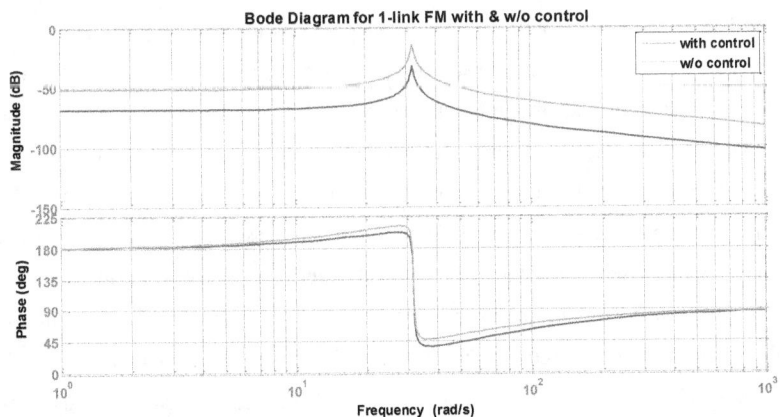

Fig. 5.5 : Bode diagram of a 1-link flexible manipulator to show effectiveness before & after the FOS control

5.3 Control design for dual link flexible manipulator

It should be noted that in this case, a dual / double / 2-flexible links are connected to obtain the overall plant structure so that the 1st motor or the 1st actuator is connected to the hub, which is termed as the base motor or base actuator to which the link 1 is connected and to the end of the link-1, another 2nd actuator or the 2nd motor is connected, which is termed as the link-2. Refer Fig. 1.1-1.3 for the diagrammatic representation of the flexible system with its specifications in Table 1.1. Our ROI (region of interest) is the tip of the link

2, which is to be controlled. The actuator/s are actuated upon by a control effort u_1 & u_2 such that the desired set-points are reached.

In the sense, when both the link gets actuated, the motor/s as well as the flexible links starts vibrating. At the tip of the flexible link-2, one displacement sensor is placed, which is used to give the feedback to the system and could be used for tracking purpose. Also, at the tip of the link-1 another displacement sensor-1 is attached for feedback purposes. This displacement $x_1(\theta)$ of the link 1 & $x_2(\theta)$ of the link 2 could be considered as the state variables, which is used for tracking and bringing back to the desired position in no time as a result of which the entire system becomes a 2 input, 2 output system or a MIMO system. In other words, there are two actuators, viz., (motor 1)-base motor & (motor 2)-shoulder motor. These displacement $x_1(\theta) = y_1$ & $x_2(\theta) = y_2$ are nothing but the outputs of the 2-link flexible manipulator system.

The first task in designing FOS controller is to fix the sampling time (Δ). An external force v (voltage is applied to the actuator) for a particular duration of t sec at the hub of the flexible link & once the motor is actuated, the system starts to move & hence is subjected to vibrations and the open loop impulse response (plot of tip output y_1 as a function of t – end of link 1 & plot of tip output y_2 as a function of t – end of link 2) of the plant is observed. The maximum bandwidth for all the tip displacement sensor & motor-actuator locations (at the base) on the flexible system are calculated (here, the second mode of the overall flexible plant) and then by using the existing empirical rules for selecting the sampling interval based on bandwidth, approximately 10 times of the maximum second mode of the flexible plant is selected, i.e., Δ is selected as $\Delta = 0.004$ s & the number of sub-intervals N is chosen to be 10.

The FOS control law is graphically displayed in the Fig. 5.2 with its block diagrammatic approach in the Fig. 5.3. Next, the given CT system is discretized and the Δ & τ systems are obtained. It is observed that both the Δ & τ systems are controllable and observable. In our case, the controllability index is given by 4 & the observability index is also obtained as 4, in the sense that all the states are controllable & observable. Next, the stabilizing output injection gains L are obtained such that the Eigen values of the Δ - DT system is inside the unit circle and is stable.

Next the τ - DT system is considered and the FOS gain **L** is found out be solving the control law equation after taking the sensor output, followed by the LMI optimization concepts such that we get an optimal value of the FOS gain so that it reduces the amplitude of the control effort u before taking the end of the flexible link to the desired position. The closed loop response, i.e., the displacement y_1 (θ) ... end of link 1 & the displacement y_2 (θ) ... end of link 2 is observed. It could be observed that the closed loop response settles at a faster rate compared to the open loop response.

The designed fast output sampling feedback controller is put in the loop with the simulated plant and the closed loop step response, i.e. response and variation of the control effort 'u_1' & 'u_2' with time 't' is observed. We need 2 control efforts as there are 2 actuators or 2 motors to control the 2 links & take it to the desired position. The state feedback gain matrix F_1 & F_2 is obtained by using the pole placement technique (place command in Matlab). $F = [F_1\ F_2]^T$ is obtained such that the Eigen values of the closed loop system $(\Phi_\tau + \Gamma_\tau F)$ lies inside the unit circle and the response of the system has a good settling time, i.e., stabilizing state feedback gains are obtained for the tau systems such that the eigenvalues of $(\Phi_\tau + \Gamma_\tau F)$ are inside unit circle. It should be noted that the effect of the state feedback is realized by a multirate output feedback gain **L**. The State Feedback Gain (SFG), F is obtained as

$$F = \begin{bmatrix} -0.2134 & 0.3354 & -0.4659 & 0.5758 \\ 0.3534 & -0.2314 & 0.5578 & -0.4569 \end{bmatrix} \quad (5.24)$$

The closed loop impulse response of the 2-link flexible manipulator system with the state feedback gain F is also observed. The obtained state feedback gain F is further used to design the FOS controller for the original system. Then, the FOS gain **L** can be determined by solving the gain equation $\mathbf{LC} = F$. But, the magnitude of this FOS gain obtained **L** is very large and hence Actually, the gains obtained by the exact solutions are very large and they may cause large variation of inputs during an output sampling instants, which may also require more control effort u exceeding the hardware limitations and cause more SNR to occur.

$$\mathbf{L} = \begin{bmatrix} -10.7134 & 9.3358 & -10.4699 & 8.5658 \\ 9.9534 & -10.2319 & 8.5568 & -10.4669 \end{bmatrix} = \begin{bmatrix} \mathbf{L}_1 & \mathbf{L}_2 \end{bmatrix} \quad (5.25)$$

Moreover with large feedback gains, the flexible system becomes more sensitive to noise and saturation may occur. So, an $L\ M\ I$ optimization procedure is used to get the optimized value of the FOS gain **L**. The value of the performance matrices in the $L\ M\ I$ concepts, i.e., ρ_1, ρ_2 & ρ_3 are tuned properly to get the optimized FOS gain vector **L**, where the gains are having less values so that input saturation will not occur. This gain vector **L** is having lesser magnitude & further it is found that the closed loop response of the original system with FOS controller gain **L** gives a stable and desired optimal behavior.

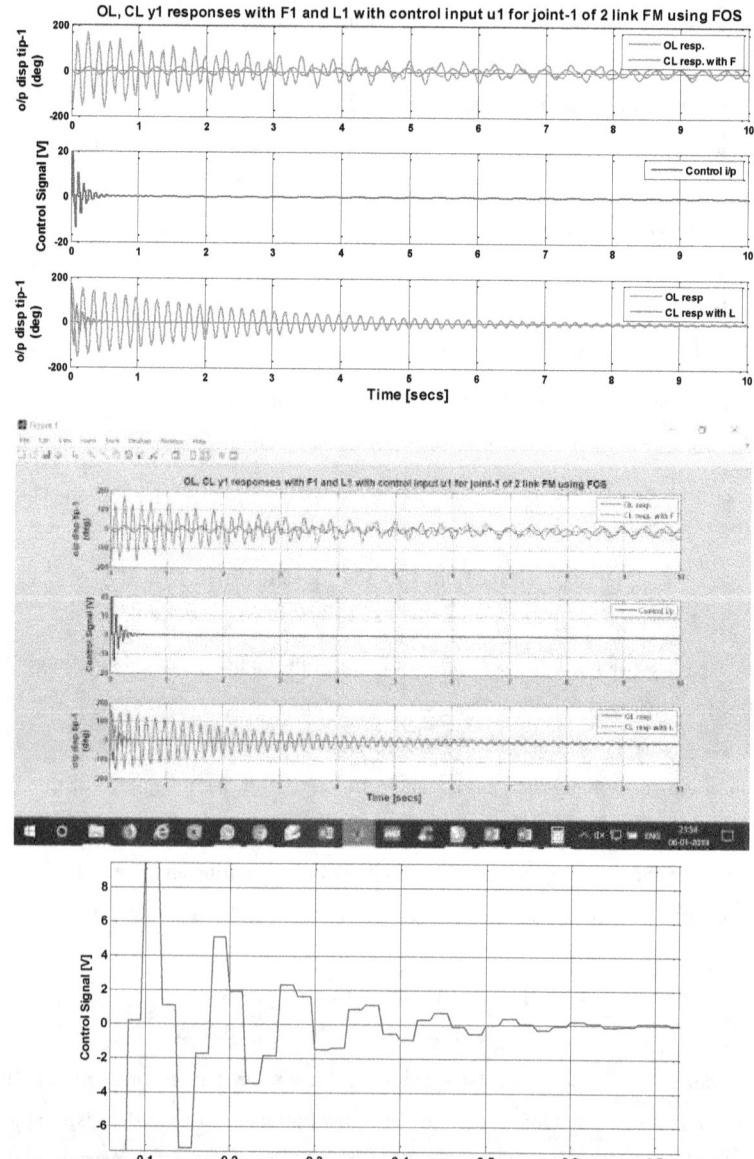

Fig. 5.6 : Open loop & closed loop response of a 2-link flexible robotic manipulator with SFB gain F_1 & FOS gain L_1 with control input u_1, output displacement of tip-1, end of link-1 (shoulder joint), y_1 & the zoomed version of the control i/p u.

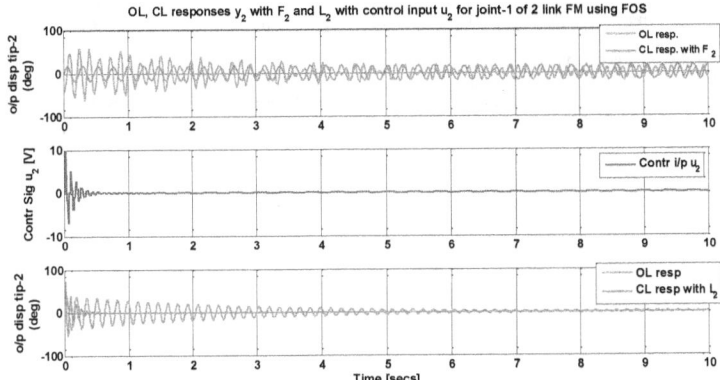

Fig. 5.7 : Open loop & closed loop response of a 2-link flexible robotic manipulator with SFB gain F_2 & FOS gain L_2 with control input u_2, output displacement of tip-2, end of link-2 y_2

These FOS gains **L** are of much smaller magnitude than the FOS gains obtained before the optimization procedure. Obviously, the control effort u is very much reduced and the variation of input during an output sampling interval is also less, thus not taxing the system, further involving less computations. The FOS controller designed by the above method requires only constant gains and its implementation is very much easier. It should be noted that in the 2-link case, the **L** is a matrix of (2×4) size, C is a matrix of size (4×2), hence, F would be a matrix of (2×4) size. The frequency response plot of the 2-link FM system is shown in the Fig. 5.8.

$$\mathbf{L} = \begin{bmatrix} \mathbf{L}_1 & \mathbf{L}_2 \end{bmatrix} = \begin{bmatrix} -0.0352 & 0.1346 & -0.3756 & 0.6357 \\ 0.2436 & -0.0536 & 0.5456 & -0.7357 \end{bmatrix} \quad (5.26)$$

The value of ρ_1, ρ_2 & ρ_3 chosen (tuned) to obtain the optimal response in the performance index calculation of *norms* of ρ was $\rho_1 = 0.02$, $\rho_2 = 0.004$ & $\rho_3 = 0.0006$ respectively. Codes are developed in the Matlab environment as .m files. The developed .m files are run, the test input is given as the input to the developed Matlab code & after running the simulation, the simulation results are observed for the contribution #3 and are presented as shown in the Figs. 5.6 & 5.7 respectively for the single link flexible manipulator case. Also, the bode plot in the frequency domain with & without the control effort is also shown here in this context in Fig. 5.8 for the sake of authentication to show the control effect. It has to be noted in this case that the gains are of dimensions (2×4) as it is a MIMO case (2 input 2 output as there are 2 motors & 2 output displacements), but the state space model (A) is of (4 × 4) and only input and the output vectors get changed in the state space model.

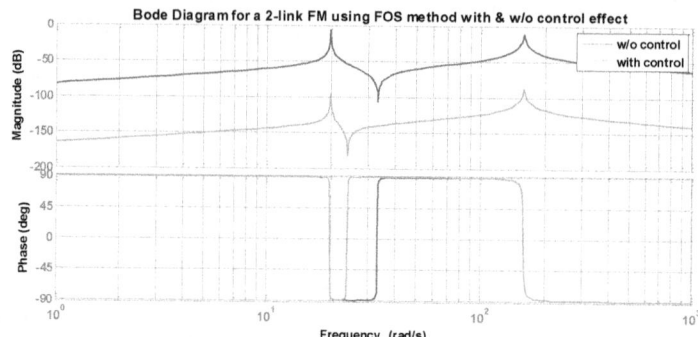

Fig. 5.8 : Bode diagram of a 2-link flexible manipulator system to show effectiveness before & after the FOS control

Till the previous cases, the simulations were carried out for a stepped excitation, in the next case, simulation is carried out for a sinusoidal excitation & the same designed FOS controller is used to get the output responses of the link-1 & link-2 (end-point displacements) of a 2-link flexible robotic manipulator. The sim results are shown in the Figs. 5.9 & 5.10 respectively for the 2-link FM case. From these sim results, it could be inferred that the magnitude of the control effort needed to control the link-2 (outer ±10 V) is lesser compared to that of the link-1 (inner ±20 V) as the base has to bear the entire weight of the flexible system.

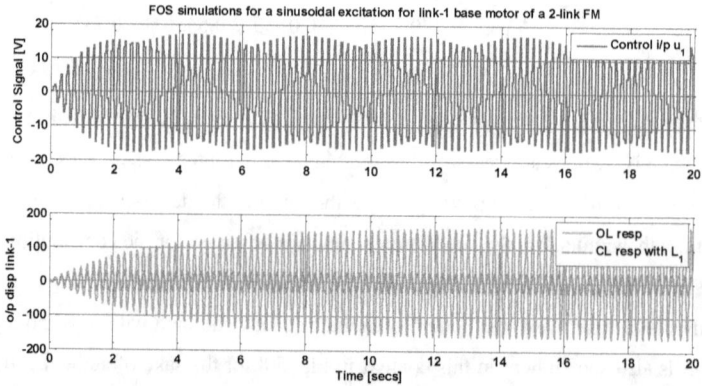

Fig. 5.9 : OL & CL response of the link-1 of a 2-link FRM with FOS gain L_1 & control u_1

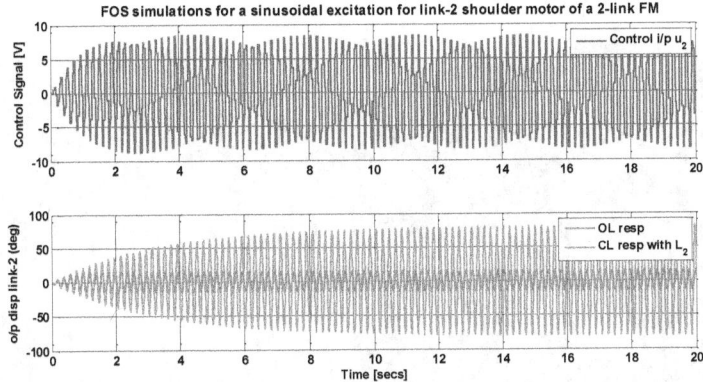

Fig. 5.10 : OL & CL response of the link-2 of a 2-link FRM with FOS gain L_2 & control u_2

5.4 Development of the FOS simulink model

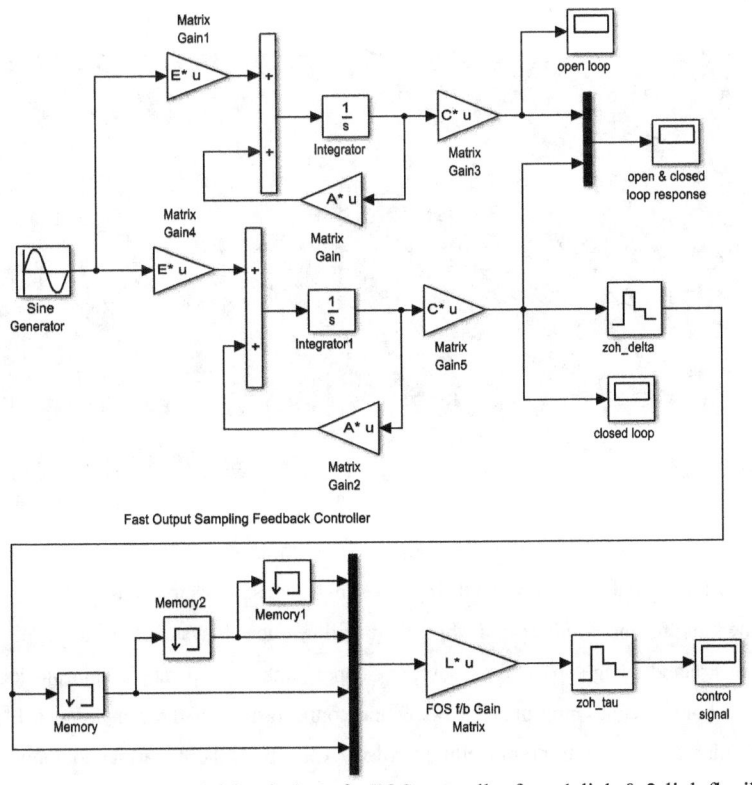

Fig. 5.11 : Simulink model for design of a FOS controller for a 1-link & 2-link flexible robotic manipulator system

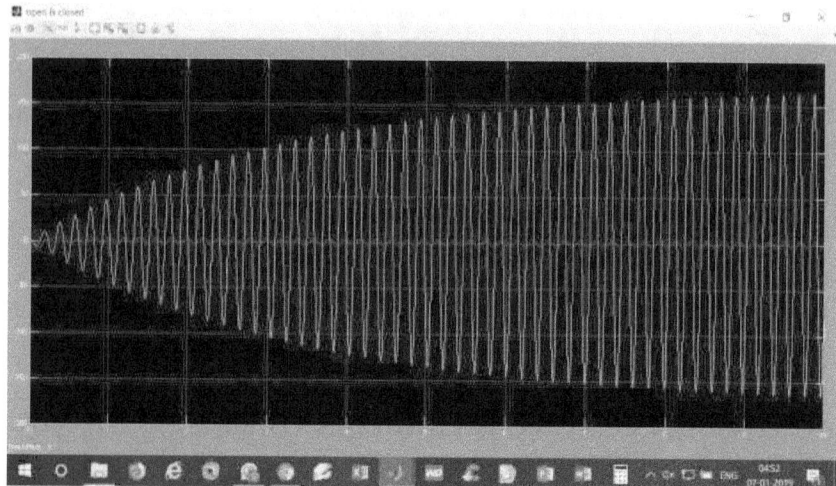

Fig. 5.12 : Scope results of OL & CL for the base motor (link-1) … amplitude more

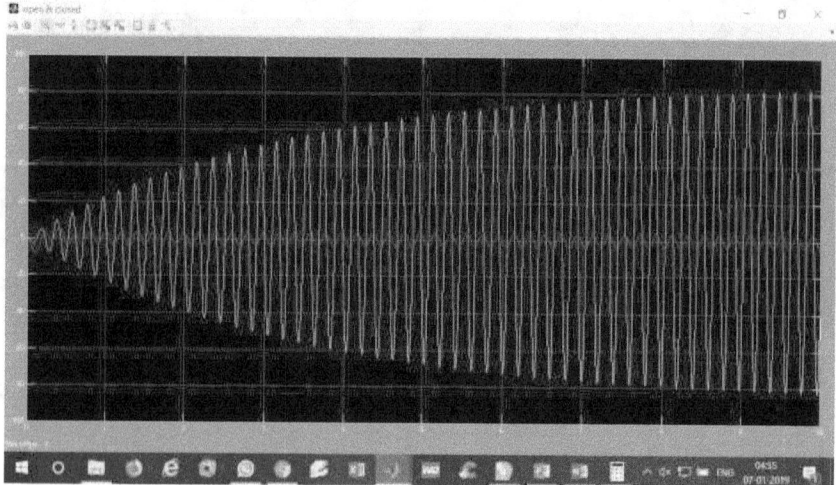

Fig. 5.13 : Scope results of OL & CL for the shoulder motor (link-2) … amplitude less

The FOS controller is also developed in the simulink environment as shown in the Fig. 5.11 for a 2-link case in which there are 2 sub-systems. It has to be noted that the model is the same for the 1-link case in which there will be one sub-system. The simulink model is constructed using sub-systems, sources, scopes, sinks, comparators, gain blocks, sample and hold circuits, multiplier blocks & the connectors. All these mentioned blocks are available in the simulink modelling library. Apart from these, various toolboxes such as control system tool box, optimization tool box, signal processing tool boxes available in the simulink library is being used. Various parameters are to be set in the different blocks that are used in the development of the simulink model.

Simulation time taken is 0 to 10 seconds (varying-any time can be taken). For the discretization purposes, the time, i.e., sample time used in the simulation is 1 ms. The developed simulink model is the same for both the types of flexible systems & only the number of sub-systems will be different. The model is run for the given requisite simulation time & the same results obtained as shown in the Matlab Simulation outputs is observed for a 2-link case & the results are not shown here for a 1-link case here for the sake of convenience. For simplicity, the scope outputs w.r.t. the links 1 & 2 is shown in the Figs. 5.12 & 5.13 respectively.

5.5 Conclusions

Research was conducted on the set-point control, end-point displacements, control of the joints of the tip-1 & tip-2 of the single link and double link flexible robotic manipulator. CT state space model of the 1 & 2 link system was developed. The CT state space model was discretized @ tau & delta sampling intervals and the discretized systems was obtained for developing the controller. Fast Output Sampling Feedback controllers were designed for the flexible manipulators to control the joint 1 of –link 1 of 1-link manipulator & the joints 1 & 2 of – the 2 link flexible manipulator. The various responses are obtained for each of the state space models of the 2 individual systems.

The developed .m file was run & the simulation results were observed. Through the simulation results, it is inferred that when the plant is placed with this designed controller, the plant performs well, the output tip displacements reaches the set-point quickly in the form of the closed loop response performing much better than the open loop response. From the responses of the double link flexible manipulator, it is observed that the output displacement of the link-1 is more compared to that of the 2-link case as it has to drive the link-1 & 2 *plus* the motor-2 weight also along with the end-effector payload, which is housed @ the shoulder level point. It has to be noted that the flexible manipulator taken into consideration is a planar 2-link one which moves in the $x - y$ plane.

In the 1-link case, the control effort is ±5 units, whereas in the 2-link case, the effort is ±20 units for the same base motor. It could be also noted that in the 2-link case, the control effort required for the base motor is ±20 units as it has to drive a less load (only the link-2) *plus* the end-effort mass of the payload, whereas for the link-1 it is ±10 units. It can also be noted that control effort u required from the controller gets reduced if the actuator is moved away from the fixed end to the end-effector point. A small magnitude of the control signal is sufficient to control the joint-2 of the 2-link case (±10 units).

A state feedback gain for each discrete models (tau & delta systems) of the 2 flexible systems is obtained such that poles are placed inside the unit circle at appropriate locations and the CL flexible system has a very good settling time. A very small magnitude of control input u is required to control the actuator as it is moved away from the base in a 2-link case as a result of which less effort has to be put by the controller (since the link-2 motor has to take care of only the link-2, whereas the link-1 motor has to bear both the link-1 & link-2 motor). The magnitude of the impulse response (closed loop) of both the continuous and the fictitious system (intermediate, i.e., with SFB gain F) is less compared to their open loop counterparts.

The close loop response characteristics with F and **L** are also the best & settle out quickly. Thus, the observations are made with and without the controller to show the control effect. From the simulations, it was observed that without control the transient response was unsatisfactory, takes more time to settle and with control, the closed loop response is. Unlike static output feedback, the FOS feedback control scheme always guarantees the stability of the closed loop system which can be observed from the simulation results.

The designed FOS controller requires constant gains and hence may be easier to implement in real time application. The controllability & observability index obtained is 4 so that all the states are controllable & observable. One excellent advantage of the proposed methodology #3 developed is that the computation time required for processing & getting the output is just within 3-4 seconds, which shows the advantage of our proposed method over the others [91] & [97]. One advantage of the 2-link mechanisms is that it can cover a greater area of the work-space & the tip or the end-effector can be tracked in a bigger dimension, i.e., R^2 area as the flexible manipulator works in the planar environment ($x - y$ place).

The work is carried out for both the step & sinusoidal excitations & here only the Matlab & Simulink's simulation results for step excitation is shown for the sake of convenience.

Chapter – 6

Design of Controllers using Discrete Sliding Mode Control for 1 & 2-link FMs with output samples

In this 4th contributory chapter, a brief *introduction* about the controller design using the discrete sliding mode control theory from the output samples is dealt with aftermath being used to control the various parameters of the 1-link & 2-link flexible manipulator. Once, the controller is designed using the sliding mode concepts, the developed controller is put in loop with the plant (1-link or 2-link flexible manipulator) and the control strategy developed is tested for its effectiveness. Matlab tool is being used to develop the control algorithm, apply it to the flexible system and then to observe the simulation results. The research work done is compared with some of the works of the yesteryears [91] & [97]. The simulink model for the DSMC designed controller using the output samples is also developed & the simulation results are observed for both the cases of the flexible manipulator. The chapter concludes with the discussion of the simulation results followed by the overall conclusions of the chapter no. #6.

6.1 A brief insight into the development of the DSM Control Design using o/p Samples for FMs

Variable structure systems exhibiting sliding mode motion has been investigated since long from the times of Gao, who was the pioneer in this field. Because of the simple structure of the sliding mode controller, they are used in different applications such as power converter, disk drives servos, robot manipulator, induction motor, stepper motor control and for a host of automobile control applications, etc. Due of the advent of computers, this computer implementation of sliding mode control has motivated the research with respect to the discrete sliding mode control algorithms. Discrete sliding mode control has been investigated since 1950's. The switching frequency cannot exceed the sampling frequency because the control signal u is always held constant between the consecutive sampling intervals. Because of this limitation, it results in chattering in discrete sliding mode control. Hence, a concept known as quasi sliding mode became popular due to the chattering effect (which is a zig zag like motion). The sliding mode, discrete sliding mode, band, chattering, etc. can be best understood from the Fig. 6.1.

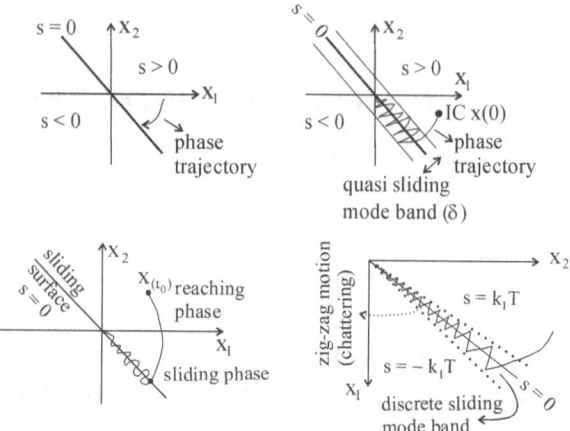

Fig. 6.1 : Sliding mode control & the discrete sliding mode control

The chattering phenomenon, which is the inherent property of any system is always present & cannot be avoided, but can be limited. The important point is that in quasi sliding mode the sliding mode motion cannot be confined on the sliding manifold and hence the invariance properties found in continuous time sliding mode does not hold for its discrete counter-part. Utkin says, in case of discrete variable structure system, the quasi sliding mode motion is possible as control action can be activated only at sampling instants. This inherently introduces a chattering known as discretization chatter and the system motion confines to a band known as the quasi sliding mode band as shown in the Fig. 6.1.

Gao proposed a method for discrete sliding mode control using reaching law approach. This reaching law assures that the system state trajectory will hit the switching manifold & undergo a zigzag like motion, thus resulting in a quasi-sliding mode motion, about the switching manifold and finally remains within a band called as the '*Quasi Sliding Mode Band (QSMB)*'. According to the concept proposed by Utkin and Drakunov, the system shows sliding mode behavior if their motion is confined to a switching manifold in the state space that is going to be reached in a finite time. Then, a discrete time sliding mode control may arise in the system with piecewise constant control u. In this context, an important and interesting possibility for discrete processing in discrete sliding mode had been proposed by Milosav in his research paper.

Adaptive sliding mode control for DT-LTI which guarantees that the ideal sliding begins after a finite time interval was presented by Utkin in a number of research papers

& used it for a variety of applicational problems. Robust DTSMC based on the above control was been proposed by Dra, which showed robustness to parametric variations. Conditions for existence of discrete time sliding mode control and control law which eliminates the zigzag motion of the discrete time sliding mode was been presented in Yu.

Another robust DT chattering free algorithm was proposed by Milosevic, which showed that the state trajectory reaches in the vicinity of the sliding manifold in finite time using the non-linear control & then it reaches to the sliding manifold in one single step using an appropriate linear control law $u(k)$. Hence, discrete sliding mode motion without chattering was obtained. It was shown that in the presence of the disturbances, the system motion confines to a small band around the sliding manifold.

Most of the sliding mode control problems require a full-state feedback. But in practical & some ideal situations, the measurement of all the system states might be neither possible nor feasible, i.e., cannot be fully measurable. Since the system output is available, output feedback can be used to design the controller as output of the system is always available as once you give the input, we get the output, so, make use of this output y. Recently DSMC & robust DSMC which results in quasi sliding mode motion was proposed by couple of authors using fast output sampling feedback techniques, which used the output of the system y.

The famous four authors DTSMC algorithm proposed by Thakkar, Saaj, Janardhanan & Bandyopadhyay [101] on the stepper motor controller design using the multirate output feedback sliding mode control concept has been used by many researchers across the wold for various applications. In our research work considered, the famous 4 author algo [101] is being used to control the 1-link & 2-link flexible robotic manipulator for the end-effector precise control. Also, the authors had presented a chattering reduction algorithm which is accomplished using the DT power rate reaching law and the concepts of MROF-multirate output feedback.

It was shown that with appropriate choice of controller parameter, the fast reaching to the set state with less chattering can be obtained& thus provided a great breakthrough in the sliding mode approach. In another algorithm which they had presented, a reaching law was modified by developing a controller having 2 parts, viz., a non-linear part and a linear part. If the non-linear parts brought the state trajectory into the vicinity of the sliding manifold & the linear part used to bring the sliding motion along the sliding manifold just in one step as a result of which the chattering was removed to a maximum extent. The authors also discussed the issues of uncertainties by taking a number of numerical examples of state space models of different sizes.

Control of flexible manipulators finds a lot of applications in the modern day world, especially in the field of avionics, robotics & the smart intelligent flexible systems. The need for a sophisticated control arises from the discrete sliding mode concepts, which has gained a lot of advantages over the traditional control schemes for controlling the various parameters of the plants. One such sliding mode control algorithm is the discrete sliding mode control abbreviated as DSMC. The output of the plant, i.e., the *'displacement'*, velocity, acceleration (each one obtained by the differentiation of the earlier ones) is considered as the parameter for control purposes, in our case, the displacement is being considered as the control variable to be tracked upon with. A brief review of which follows in the following paragraphs.

This special discrete control algorithms' developed to cater for the flexibility in the robotic arms and these algorithms are not that complex, considering that the system is non-linear as generally all flexible systems even though they are non-linear will be linearized about an operating point.

In the research work considered, DSMC is used to regulate the motion of the flexible manipulators & take the tip to a specified location and also to track the set-points, curb down the vibrations with less errors. To design the controller, a state space model is obtained first from the first principles starting from the **M**, **B**, **K** parameters. 2 separate cases of the flexible robotic manipulator are presented here with, viz., the SISO (1-link) control, second-the MIMO (2-link) control, in the sense the former has one actuator – so, 1 input & 1 output, whereas the latter has two actuators – so, 2 inputs & 2 outputs. Open loop (OL) & closed loop (CL) responses are also observed with the gain of the DSMC. Simulations are carried out with & without the DSMC controller to show the authenticity of the proposed control strategies in comparision with the work done by other researchers [91] & [97].

The DSMC control algorithm actually needs the plant system to be completely controllable & be observable, thus giving the desired performances even in the presence of un-measured disturbances. The performance of the DSMC controller will be tested on the 1 & 2 link flexible manipulator & could be generalized for a n-link flexible manipulator case also, which could be treated as *future works*.

The Sliding Mode Control (SMC) theory is actually developed on the concept of changing the structure of the controller in response to the changing state x of any plant or the system such that the desired response y is obtained, the SMC theory being proposed by Utkin. Hung & Young said, "*a high speed switching control action can be used in order to switch between different structures & the trajectory of the given system can be forced to move along a chosen manifold in the state space*". This being called as the switching manifold. The behaviour of the closed loop control system is therefore determined by using the sliding surface. Till date, a large number of researchers have put forth lot of efforts in studying the Discrete Sliding Mode (DSM) and quite a number of controller designs have been put forward by many authors & all those control laws were used for a variety of applications, to name a few of them, Gao, Furuta, Werner, Bag.

If the discrete sliding mode is considered, the control action is activated at sampling instants &u, the control effort is treated as constant over each sampling periodτ. When the state x reaches the switching surface, the subsequent discrete time switchings cannot maintain the state on the sliding surface. Since the DSM undergoes only quasi-sliding motion, the famous sliding mode specialist, Gao introduced a new law called as the '*reaching law approach*' to design the controller for any discrete-time system using state feedback. This reaching law totally ensured that the system trajectory will hit the switching manifold & thereafter will undergo a zigzag motion about the switching manifold and reach the equilibrium point.

The magnitude of each of the successive zig-zagging step goes on decreasing so that the state trajectory stays within a specified band. This specified band is called the quasi-sliding mode band. Many of the sliding mode control methods requires full state feedback (SFB). But in some of the ideal & practical situations, it is a well-known fact that all the states of the system may not be available for measurement and hence the control may or may not be feasible.

Since the system output is always available (*any system has 3 important parts – input, system, output*), the output of the system can be feedback & this output y can be used to design the controller to control a particular variable or a group of variables, say θ, ω, & α, i.e., angular displacement, angular velocity and angular acceleration.

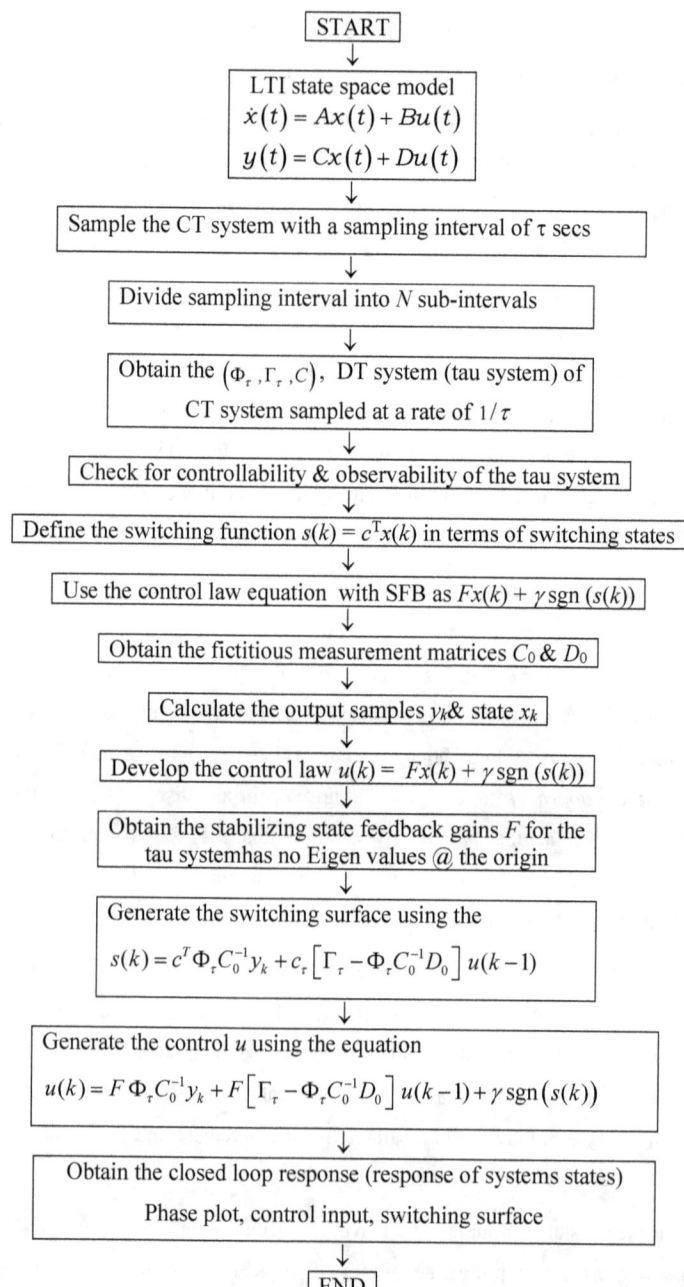

Fig. 6.2 : Proposed flow-chart for the controller design using the DSMC concepts

Considerable research has been carried out in the field of output feedback sliding mode control. The static output feedback (SOFB) problem is one of the most investigated problems in majority of the control theory & its applications. But, still now, no results are available till today which has shown that the complete pole assignment is possible using static output feedback. Because of this reason, one can resort to output feedback as the output of the system is readily available.

In this context, the output feedback can be realized using fast output sampling concepts developed by Werner & Furuta. Werner has used the fast output sampling (FOS) feedback which has the features of static output feedback and makes it possible to arbitrarily assign the system poles. Unlike static output feedback, fast output sampling feedback always guarantees the stability of the closed loop system. Recently a fast output sampling sliding mode control algorithm (FOSSMC) was proposed by Saaj, which uses reaching law approach. The output is corrected before being used for feedback purpose.

The output feedback gain L is obtained from the state feedback gain F using the relation $LC = F$ proposed by Werner. An improvement of the technique was proposed by Saaj, where output correction is not required to generate the control signal u. The method proposed by Saaj does not need the calculation of the output feedback gain L using the previously mentioned relation. In this context, the switching function $s(k)$ and control input u is directly obtained in terms of the immediate past control values u and the past output samples y.

The proposed flow-chart for the controller design using the discrete sliding mode concept using output samples is depicted in the Fig. 6.2. Now, to implement this proposed flow-chart shown in the Fig. 6.2 for the controller design using the DSMC concepts on the 1-link & 2-link flexible robotic manipulator, we consider a plant described by a LTI state space model given by the equation (6.1) as

$$\dot{x}(t) = Ax(t) + Bu(t)$$
$$y(t) = Cx(t) + Du(t) \quad (6.1)$$

where $x \in R^n, u \in R^m, y \in R^p, A \in R^{n \times n}, B \in R^{n \times m}, C \in R^{p \times n}$, A, B, C are constant matrices and it is assumed that (A, B) is controllable and (C, A) is observable. It is to be noted that A, B, C, D, x, u, y are the system matrix, input matrix, output matrix, transmission matrix, state variable, input variable & the output variable of the plant (1-link & 2-link flexible system) & are constant matrices of appropriate dimensions.

Consider the discrete system obtained using the sampling of the CT system by τ units

$$x(k+1) = \Phi_\tau x(k) + \Gamma_\tau u(k) \quad \&$$
$$y(k) = C\,x(k) \tag{6.2}$$

where τ is sampling period, x is n-dimensional state vector, u is scalar, y is output vector, & the matrices Φ_τ, Γ_τ and C are of appropriate dimensions. Here, we make an assumption that the matrix pairs (Φ_τ, Γ_τ) & (Φ_τ, C) are controllable & observable. A reaching law for the SMC of a DT system as proposed by Gao, which has the equation of the following form & is given by

$$s(k+1) - s(k) = -q\,\tau\,s(k) - \varepsilon\,\tau\,\mathrm{sgn}(s(k)) \tag{6.3}$$

where $\tau > 0$, is the sampling period, $\varepsilon > 0$, $q > 0$, $(1 - q\tau) > 0$ and $s(k)$ is the switching function defined as a function of the system states as

$$s(k) = c^T x(k) = 0 \tag{6.4}$$

Using the reaching law given in the equation (6.3), the control law $u(k)$ for the system represented by the equation (6.2) has been obtained as per Gao with the control model

$$u(k) = F\,x(k) + \gamma\,\mathrm{sgn}(s(k)) \tag{6.5}$$

where

$$F = -\left(c^T \Gamma_\tau\right)^{-1}\left[c^T \Phi_\tau - c^T I + q\,\tau\,c^T\right] \tag{6.6}$$

$$\gamma = -\left(c^T \Gamma_\tau\right)^{-1} \varepsilon\,\tau \tag{6.7}$$

Bartoszewicz proposed the quasi-sliding mode band's width δ as

$$2\delta \leq \frac{2\varepsilon\tau}{2 - q\tau}. \tag{6.8}$$

It is to be noted that the system states like within the steady state band given by the above band's width δ equation.

Fast Output Sampling (FOS) technique was dealt with in brief in chapter 5. In this method, the output feedback law is used to realize a discrete state feedback gain by multirate observations of the output signal & the control signal is held constant during each sampling interval τ & this concept is being clubbed with the discrete sliding mode control to obtain a new type of control called as the FOS-SMC with output samples as proposed by the famous 4-author algorithm (Thakkar, Saaj, Janardhanan, Bandyopadhyay), a small review of which is presented as follows [101].

Here, the reaching law approach as proposed by Gao, where the control takes the form as mentioned in equation (6.5) is being used with the value of the γ as defined in the equation (6.7). Next the control law $u(k)$ is obtained as follows after applying to the discretized system in equation (6.2) as

$$x(k+1) = (\Phi_\tau + \Gamma_\tau F)x(k) + \Gamma_\tau \gamma \qquad (6.9)$$

$$y_{k+1} = (C_0 + D_0 F)x(k) + D_0 \gamma \qquad (6.10)$$

Using the equation no. (6.9), it can be deduced further as

$$x(k) = (\Phi_\tau + \Gamma_\tau F)^{-1} x(k+1) - (\Phi_\tau + \Gamma_\tau F)^{-1} \Gamma_\tau \gamma \qquad (6.11)$$

Now from the equation (6.11), substitute for $x(k)$ from equation (6.10) & further simplifying all those relations, we get the relation between the states & the output samples y_k as and simplifying them gives the relation between output samples $x(k)$ and y_k and state $x(k)$ as

$$y_k = Cx(k) + \alpha \quad \& \quad \alpha = (D_0 - C\Gamma_\tau)\gamma \qquad (6.12)$$

When $x(k)$ is substituted from the equation (6.12) into equation (6.5), we get the output feedback control law, which does not need the output correction as given by Saaj & Bandyopadhyay as given by the equation (6.13) in

$$u(k) = L y_k + \eta(k); \quad \text{where} \quad \eta(k) = -L\alpha + \gamma \qquad (6.13)$$

The above mentioned control u drives the system states x towards the sliding manifold & once the state trajectory crosses the switching line, the control u will assume the form

$$u(k) = L y_k + \eta(k) \qquad (6.14)$$

where $\eta(k)$ switches between the values given by $\eta_1 = L\alpha + \gamma$ and $\eta_2 = -(L\alpha + \gamma)$ at every subsequent sampling instant τ. Finally, the relation for switching function $s(k)$ in terms of the output is given by

$$s(k) = c^T \mathbf{C}^{-1}(y_k - \alpha) \tag{6.15}$$

Now, in order to determine the mathematical model for the SMC using the output samples, consider the discretized CT systemby τ, i.e., the tau system with the value of $k = 0$ as given by the equation (6.16) as

$$x(\tau) = \Phi_\tau x(0) + \Gamma_\tau x(0) \quad \& \quad y_\tau = C_0 x(0) + D_0 u(0) \tag{6.16}$$

Now, at this juncture, the output samples are not known before the time of $t = 0$ & therefore the control signal can be deduced using the equation (6.5) by putting $k = 0$, thus the switching can be deduced as finally

$$s(\tau) = c^T \Phi_\tau C_0^{-1} y_\tau + c^T \left(\Gamma_\tau - \Phi_\tau C_0^{-1} D_0 \right) u(0) \tag{6.17}$$

& the control at τ is given by

$$u(\tau) = F \Phi_\tau C_0^{-1} y_\tau + F \left(\Gamma_\tau - \Phi_\tau C_0^{-1} D_0 \right) u(0) + \gamma \operatorname{sgn}\left[s(\tau) \right] \tag{6.18}$$

Since it is assumed that $n = N$ as the system is completely observable, i.e., C_0^{-1} exists. Finally, after simplification, a generalized expression for the switching function & control as

$$s(k) = c^T \Phi_\tau C_0^{-1} y_k + c^T \left(\Gamma_\tau - \Phi_\tau C_0^{-1} D_0 \right) u(k-1) \tag{6.19}$$

$$u(k) = F \Phi_\tau C_0^{-1} y_k + F \left(\Gamma_\tau - \Phi_\tau C_0^{-1} D_0 \right) u(k-1) + \gamma \operatorname{sgn}\left(s(k) \right) \tag{6.20}$$

From the above equation of $u(k)$, it is observed that the system states are needed neither for the switching function evaluation nor for any feedback purposes. This concept is being used in our work to control the 1-link & 2-link flexible robotic manipulator's some of the system parameters such as the position, etc.

6.2 Control design for Single link flexible manipulator

It should be noted that in this case, a single flexible link is connected to the hub where a single actuator is used for actuation purpose. Refer Figs. 1.1-1.3 for the diagrammatic representation of the flexible system with its specifications in Table 1.1. The actuator is actuated upon by a control effort u such that the desired set-point is reached. In the sense, when the single link gets actuated, the motor as well as the flexible link starts vibrating.

At the tip of the flexible link, one displacement sensor is placed, which is used to give the feedback to the system and could be used for tracking purpose. This displacement x (θ), which is nothing but the output of the system y could be considered as one of the state variable, which is used for tracking and bringing back to the desired position in no time. In other words, there is one actuator (motor)-input & one sensor (used to obtain the displacement of the end-effector-output as such the 1-link flexible system can be considered as a SISO system.

Simulation study has been carried out for a particular designed set values of the DSMC parameters q, τ, N, Δ, ε. The simulation results can also be observed for different sets of parameters of q, τ, N, Δ, ε & computing the switching surfaces $s(k)$ along with the control inputs $u(k)$. It can also be seen that the responses of the states x_1, x_2 converges to zero (0) from a given initial condition $x(0)$. Also, the phase plots for the flexible system shows that the system states exhibiting the quasi-sliding motion. Another response plot shows that the switching function decreases towards zero from the initial value and stay within a small band in the neighbourhood of the switching line. Since the single-link model is modelled as a (2 × 2) case, there will be only 2 states x_1 & x_2 as the size of the A matrix will be of (2 × 2). The values chosen for the tracking control were

$$q = 1, \tau = 0.2, N = 10, \Delta = 0.1, \varepsilon = 0.005.$$

From the sim results, it can be clearly observed that the set-point tracking control is good. Matlab tool has been used to develop the *.m* code. The developed algorithms are written in the form of a code, the code is run and the various responses that were considered in the sim study were observed, the results were discussed. The following parameters were observed from the simulation results.

- State trajectory
- Phase plots
- Output tip displacements
- System states
- Switching function
- Control input

From the simulation results of the various previous parameters, it can be concluded that the effectiveness of the proposed algorithms are observed as they reach the stability point in no time (just within 5 seconds). The frequency response plot also shows how the effect has taken place before & after the control effect. The control effect taken is also less as it is a 1-link manipulator case.

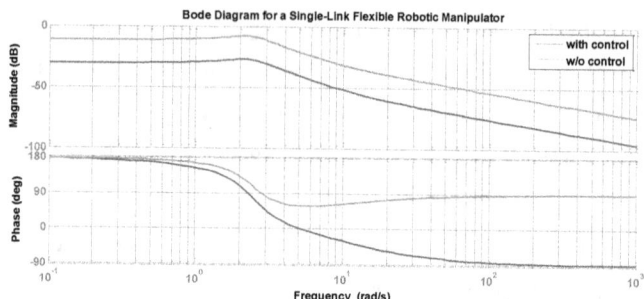

Fig. 6.3 : Bode diagram (frequency plot) for a single link-flexible robotic manipulator showing with & w/o control

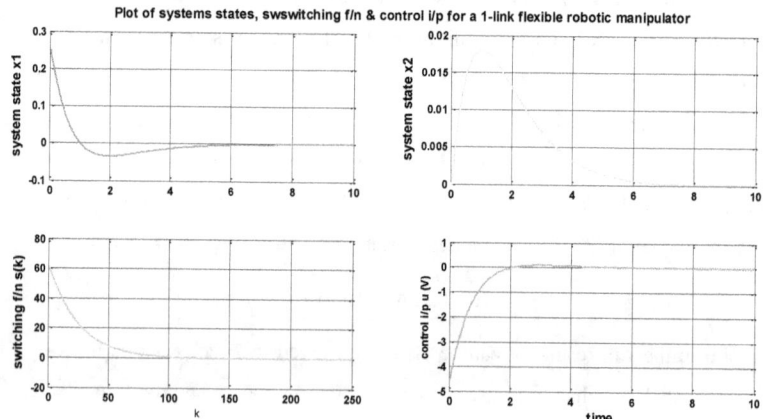

Fig. 6.4 : System responses (plot of the system states), switching function & control input for a single link flexible robotic manipulator

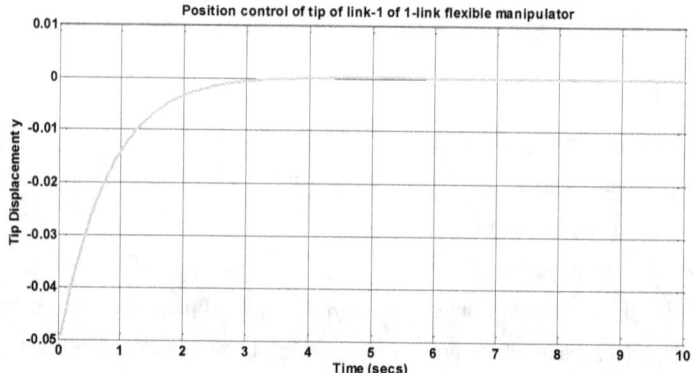

Fig. 6.5 : Tip position control setting to the track point, equilibrium point

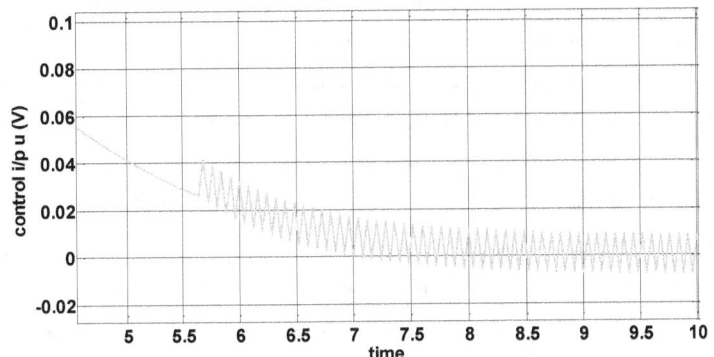

Fig. 6.6 : Control input plot showing the switching action

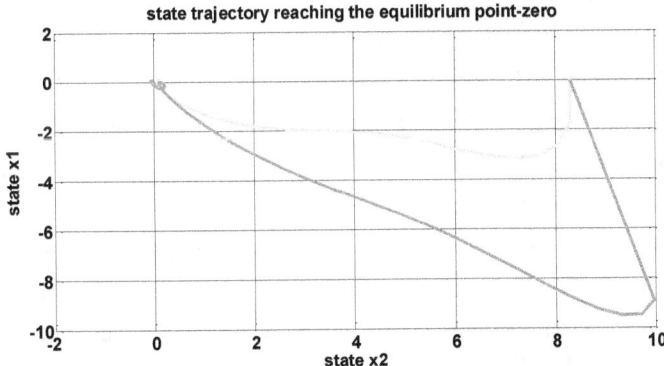

Fig. 6.7 : State trajectory (phase-plot) reaching equilibrium point of '0' showing tracking

From the Fig. 6.4/6.6, it can be seen how the chattering effect is seen, a zig-zag like motion & finally from the Fig. 6.7, how the state trajectory (phase plot) reaches the equilibrium point (origin) from a given initial condition, say $x(t_0)$ can be seen. Also, from Fig. 6.5, it can be seen how the tip is controlled to a set-point (0).

6.3 Control design for dual link flexible manipulator

It should be noted that in this case, a dual / double / 2-flexible links are connected to obtain the overall plant structure so that the 1st motor or the 1st actuator is connected to the hub, which is termed as the base motor or base actuator to which the link 1 is connected and to the end of the link-1, another 2nd actuator or the 2nd motor is connected, which is termed as the link-2. Refer Fig. 1.1-1.3 for the diagrammatic representation of the flexible system with its specifications in Table 1.1. Our ROI (region of interest) is the tip of the link 2, which is to be controlled. The actuator/s are actuated upon by a control effort u_1 & u_2 such that the desired set-points are reached.

In the sense, when both the link gets actuated, the motor/s as well as the flexible links starts vibrating. At the tip of the flexible link-2, one displacement sensor is placed, which is used to give the feedback to the system and could be used for tracking purpose. Also, at the tip of the link-1 another displacement sensor-1 is attached for feedback purposes. This displacement $x_1(\theta)$ of the link 1 & $x_2(\theta)$ of the link 2 could be considered as the state variables, which is used for tracking and bringing back to the desired position in no time as a result of which the entire system becomes a 2 input, 2 output system or a MIMO system. In other words, there are two actuators, viz., (motor 1)-base motor & (motor 2)-shoulder motor. These displacement $x_1(\theta) = y_1$ & $x_2(\theta) = y_2$ are nothing but the outputs of the 2-link flexible manipulator system.

Simulation study has been carried out for a particular designed set values of the DSMC parameters q, τ, N, Δ, ε. The simulation results can also be observed for different sets of parameters of q, τ, N, Δ, ε & computing the switching surfaces $s(k)$ along with the control inputs $u(k)$. It can also be seen that the responses of the states x_1, x_2 converges to zero (0) from a given initial condition $x(0)$. Also, the phase plots for the flexible system shows that the system states exhibiting the quasi-sliding motion. Another response plot shows that the switching function decreases towards zero from the initial value and stay within a small band in the neighbourhood of the switching line. Since the single-link model is modelled as a (4 × 4) case, there will be only 2 states x_1, x_2, x_3 & x_4 as the size of the A matrix will be of (4 × 4). The values chosen for the tracking control were

$$q = 2, \tau = 0.4, N = 20, \Delta = 0.2, \varepsilon = 0.01.$$

From the sim results, it can be clearly observed that the set-point tracking control is good. Matlab tool has been used to develop the .m code. The developed algorithms are written in the form of a code, the code is run and the various responses that were considered in the sim study were observed, the results were discussed. The following parameters were observed from the simulation results.

- State trajectory
- Phase plots
- Output tip displacements
- System states

- Switching function
- Control input

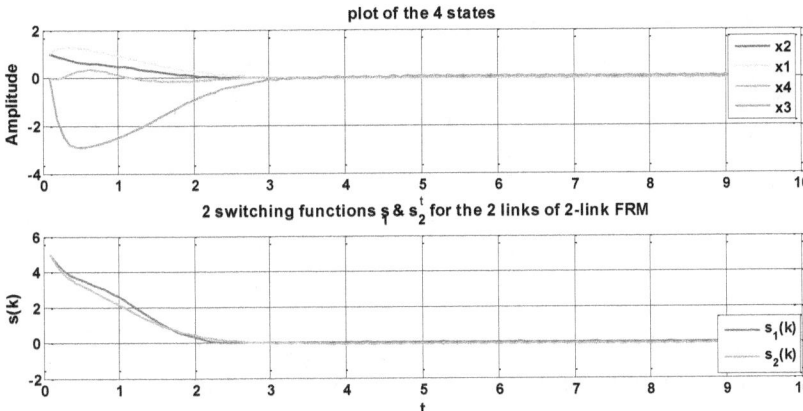

Fig. 6.8 : State plots & the switching planes of the entire 2-link combined flexible system

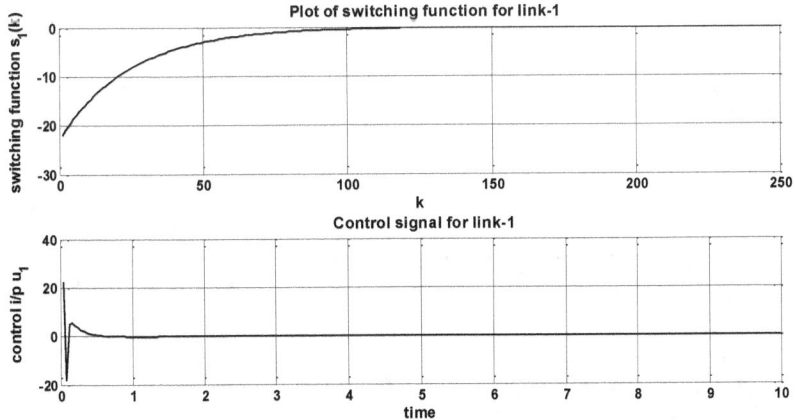

Fig. 6.9 : Plot of the control effort u showing the switching action using the switching function $s(k)$ for the single link flexible robotic manipulator

From the simulation results of the various previous parameters, it can be concluded that the effectiveness of the proposed algorithms are observed as they reach the stability point in no time (just within 5 seconds). From the frequency response plot, it can also be observed how the effect has taken place before & after the control effect (not shown here for the sake of convenience). The control effect taken is on the higher side compared to that of the 1-link manipulator case as the 2-link planar flexible manipulator has to bear the weight of link-1, link-2 & the 2 actuators.

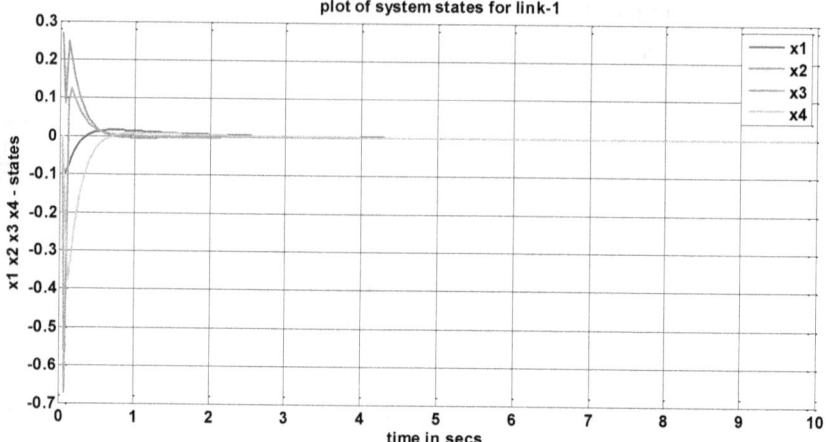

Fig. 6.10 : Plot of all the 4 system states for the link-1 of the flexible manipulator

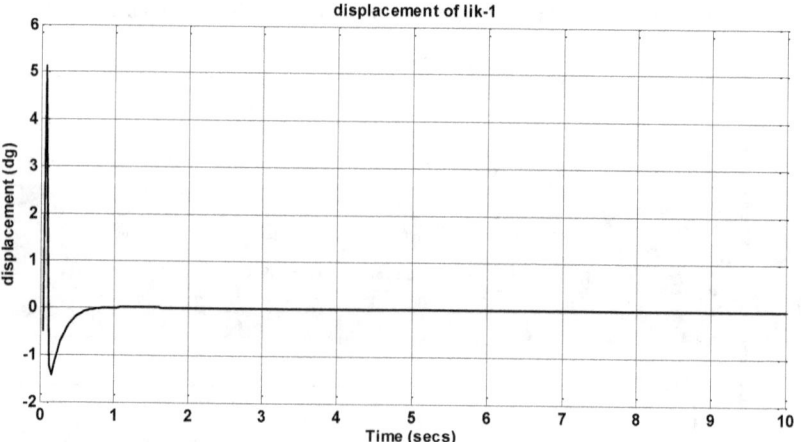

Fig. 6.11 : Tip displacement of the link-1 of the flexible manipulator

From the Fig. 6.13, it can be seen how the chattering effect is seen, a zig-zag like motion & finally from the Fig. 6.14, how the state trajectory (phase plot) reaches the equilibrium point (origin) from a given initial condition, say $x(t_0)$ can be seen. Also, from Figs. 6.11 & 6.12, it can be seen how the tip is controlled to a set-point (0) for the 2 links of the flexible manipulator. The plot of the 4 system states is also shown in the Fig. 6.10, which also shows that all the states reaches the equilibrium point.

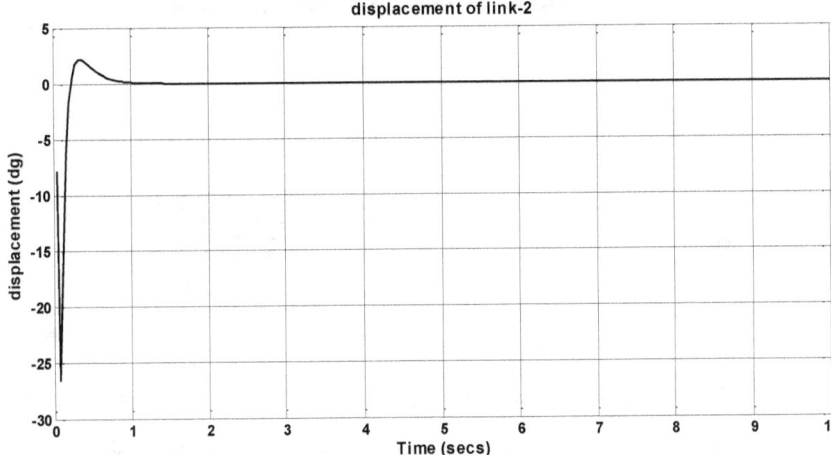

Fig. 6.12 : Tip displacement of the link-2 of the flexible manipulator

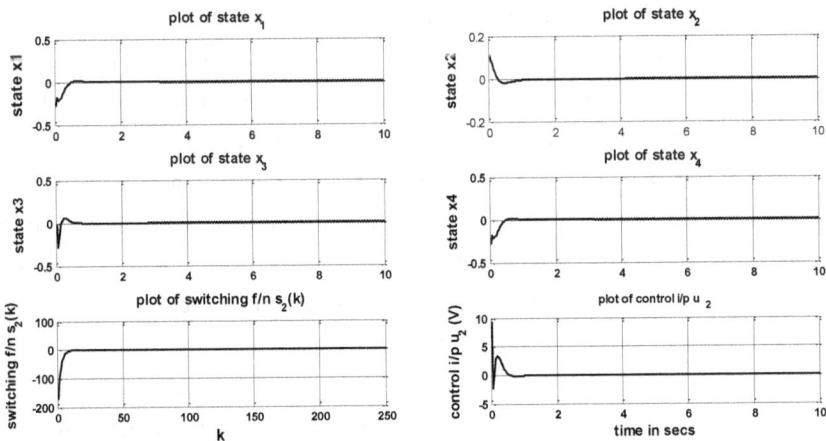

Fig. 6.13 : Plot of the system states, switching function & control input required for the link-2 of the flexible manipulator

6.4 Development of the DSMC simulink model

The DSMC controller is also developed in the simulink environment as shown in the algorithm of the flow-chart in the Fig. 6.2 for a 2-link case in which there are 2 sub-systems. It has to be noted that the model is the same for the 1-link case in which there will be one sub-system. The simulink model is constructed using sub-systems, sources, scopes, sinks, comparators, gain blocks, sample and hold circuits, multiplier blocks & the connectors. All these mentioned blocks are available in the simulink modelling library.

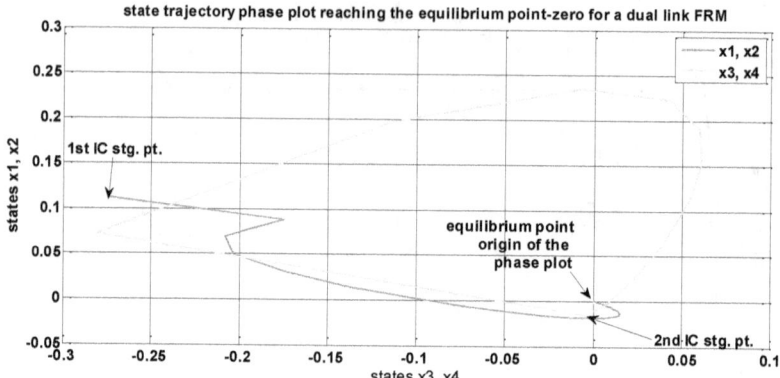

Fig. 6.14 : State trajectory (phase-plot) reaching equilibrium point of '0' showing tracking

Apart from these, various toolboxes such as control system tool box, optimization tool box, signal processing tool boxes available in the simulink library is being used. Various parameters are to be set in the different blocks that are used in the development of the simulink model. Simulation time taken is 0 to 10 seconds (varying-any time can be taken). For the discretization purposes, the time, i.e., sample time used in the simulation is 1 ms. The developed simulink model is the same for both the types of flexible systems & only the number of sub-systems will be different. The model is run for the given requisite simulation time & the same results obtained as shown in the Matlab Simulation outputs is observed for a 2-link case & the results are not shown here for a 1-link case here for the sake of convenience & is similar to the one presented in the chapter 5.

6.5 Conclusions

Research was conducted on the set-point control, end-point displacements, control of the joints of the tip-1 & tip-2 of the single link and double link flexible robotic manipulator. CT state space model of the 1 & 2 link system was developed. Discrete sliding mode controllers using output samples were designed for the flexible manipulators to control the joint 1 of − link 1 of 1-link manipulator & the joints 1 & 2 of − the 2 link flexible manipulator.

The developed .m file was run & the simulation results were observed, i.e., The various responses are obtained for each of the state space models of the 2 individual flexible systems. Through the simulation results, it is inferred that when the plant is placed with this designed controller, the plant performs well, the output tip displacements reaches the set-point quickly in the form of the closed loop response performing much better than the open loop response.

The simulation results can also be observed for different sets of parameters of q, τ, N, Δ, ε & computing the switching surfaces $s(k)$ along with the control inputs $u(k)$. It can also be seen that the responses of the states x_1, x_2 converges to zero (0) from a given initial condition $x(0)$. Also, the phase plots for the flexible system shows that the system states exhibiting the quasi-sliding motion. Another response plot shows that the switching function decreases towards zero from the initial value and stay within a small band in the neighbourhood of the switching line.

It can also be observed for a 2-link case, the control effort needed for the link-1 is more than that of the link-2 as the base actuator has to control the weight of the link-1 + shoulder motor weight + the weight of the link-2 along with its end-effector payload mass, which can be justified from the control effort required for positioning diagrams in the simulation results.

In this chapter, a new discrete sliding mode control using multirate output samples is presented for a flexible robotic manipulator. Further, it can also be observed that the DSM control law can be directly obtained in terms of the output samples y and immediate past control function u. The research work done is compared with some of the works of the yesteryears [91] & [97]. Further, one main advantage of this methodology is that the algorithm neither needs the states of the system for feedback purpose nor for the switching function evaluation & therefore can be easily implementable in real time applications as the gains are piecewise constants.

Chapter – 7

Conclusions & Future Work

In this 7th & final chapter, the brief outcome, i.e., the concluding remarks of the research work carried out along with the scope for future work in this exciting application oriented industrial research area is presented. This thesis has presented the investigations into control of a flexible manipulator system (1 & 2 link FMS). Several control approaches had initially been discussed and the research direction was accordingly identified. Simulation exercises with a single-link & a dual-link flexible manipulator with and without control have been performed. Different system responses have been obtained and presented in the time and frequency domains & finally arrived @ the following conclusions presented next.

7.1 Conclusions

Research was carried out on the control of n-link flexible robotic manipulators in the three dimensional Euclidean space. To start with, an extensive background research was carried out on the chosen research topic which was given by the supervisor. During this initial period, a large number reference books, text books, journal papers and conference papers covering the fundamental theoretical concepts which would be a background for the research work were collected & studied. Hundreds of related research papers were collected from various sources such as …. internet, library, several IITs, NITs, IISc., Foreign Universities, Deemed Universities, Different college libraries, RMYEC library, GMIT library, JSSATE library, Research Center of VTU-Belgaum, Different engineering college libraries in Karnataka & Maharashtra, VTU Belgaum Library, from friends, earlier researchers who have done similar work, etc…. This literature review was carried for a period of more than 6 months and finally summarized to get a conceptual view and was also being published as a review paper in 2 reputed IEEE conferences. This was followed by the taking of the written test, passing of the interview, taking of the 4 course work subjects & clearing them in good grades.

After collecting lot of research materials on the chosen topic and defining the problem, the problem solving methodologies of the chosen problem/s was formulated, simulations were carried out, obtained the results and the proposed works being published in reputed conferences & journals with high impact factors. A brief review of the related work bearing the essence of the control of n-link flexible robotic manipulators in the three dimensional Euclidean space was presented in the form of an abridged literature survey along with an introductory information related to the research work in the chapters 1 & 2 respectively. The objectives of the research work was also explored and arrived at the definition of the problem that had to be tackled with and solved.

A total of 4 contributory works were carried out during the course of the research work till date in the field of designing of controllers for controlling the set-points of the 1 & 2-link flexible robotic manipulators in the three dimensional Euclidean space case in 4 separate chapters, viz., chapter 3, 4, 5 & 6 respectively. The industrial specifications of the one & 2 link flexible robotic manipulator was considered for the simulation purpose. The 4 proposed works presented as 4 detailed chapters in the thesis were

1. Control of 1-link & 2-link flexible robotic manipulator using PID control scheme for controlling the set-points (position) & the vibrations of the actuators.
2. Control of 1-link & 2-link flexible robotic manipulator using Periodic Output Feedback (POF) scheme for controlling the set-points (position) & the vibrations of the actuators.
3. Control of 1-link & 2-link flexible robotic manipulator using Fast Output Sampling Feedback (FOS) scheme for controlling the set-points (position) & the vibrations of the actuators.
4. Control of 1-link & 2-link flexible robotic manipulator using Discrete Sliding Mode (DSMC) scheme for controlling the set-points (position) & the vibrations of the actuators.

The software tool that is used for the research work is Matlab 14 with Simulink & various tool boxes such as control system tool box, signal processing tool box, Optimization tool box, etc.

An automatic feedback control system to get the flexible system's output to the desired output was developed using the concepts of PID, POF, FOS & SMC methodologies. Note that to achieve this, dynamics & kinematics (forward kinematics & backward kinematics) of the robotic system was also developed for the flexible robot to achieve the target output. The output responses of the structure with & without the controller was observed, thus showing the control effectiveness and comparing of the research work done by us with other researchers in the past to establish the supremacy & effectiveness of the proposed works developed.

The research work that was undertaken by me under the guidance of my supervisor was aimed to develop sophisticated control algorithms for control of flexible manipulator systems, where flexibility of the links and the joints played an important role, that too concentrating on the tip position accuracy, trajectory control of motors, which was our main desired objective.

The final result or the outcome or the end-result of the research work was aimed @ developing some efficient control algorithms which will accurately position the tip of the end-effector in spite of all non-linearities, noises, disturbances, vibrations, etc... and to reduce the overall weight of the systems due to the flexible nature of the manipulator links, curb the vibrations / noises (unwanted signals) in just a couple of seconds using different types of sensors & actuators, adopt the hybrid type of control, i.e., position, velocity & vibration control along with motor tracking control. In short, to say, the outcome of the research is to show that when the flexible manipulator is placed with this developed robust controller, the flexible system will perform well and reaches the destination (output) in shorter lead times and will track the reference input.

Research was conducted on the set-point control, end-point displacements, control of the joints of the tip-1 & tip-2 of the single link and double link flexible robotic manipulator using PID, POF, FOS & SMC methods. CT state space model of the 1 & 2 link system was developed. PID, POF, FOS & SMC controllers were designed for the flexible manipulators to control the joint 1 of − link 1 of 1-link manipulator & the joints 1 & 2 of − the 2 link flexible manipulator.

Codes were developed in Matlab environment as .m files for all the 4 contributory works with 2 case studies in each work, viz., the single link case & the double link case. The developed codes were run, the simulation results were observed, discussions were carried out & the conclusions summarized w.r.t. each of the contributory case.

The controllers were also developed in the simulink environment for a 1-link & 2-link case as .mdl simulink models. The simulink model was constructed using sub-systems, sources, scopes, sinks, comparators, gain blocks, sample and hold circuits, multiplier blocks & the connectors, with each one taken from the simulink modelling library & finally the developed simulink model was run for a specific amount of time. The simulation results were observed, discussions were carried out & the conclusions summarized w.r.t. each of the contributory case. It was observed that the results were the same as obtained in the Matlab environment.

Through the simulation results for all the 4 developed works, it is inferred that when the plant is placed with this designed controller, the plant performs well, the output tip displacements reaches the set-point quickly in the form of the closed loop response performing much better than the open loop response in both the Matlab environment & in the Simulink environment.

To start with the 1st contribution, the research was carried out on the development of tracking control algorithms for the control of flexible robotic manipulators in the 3 dimensional Euclidean space using PID controllers first. A single link manipulator was considered followed by a 2 link manipulator for the simulation purposes as the plant model. Mathematical model of the PID was used to develop the controller to track the speed of the motors (joints in R^2 space) in the flexible link manipulators. Tuning of the controller was done using Ziegler-Nicholas method for both the case studies.

Models were developed & run to observe the simulation results and arrive at the expected output results (*goals*). The simulation results show the efficacy of the developed PID controller to control the tracking of the actuators for the single link & dual link manipulators. In the design of the PID controller, the proportional tuning involved in correcting a target proportional to the difference, the integral tuning attempted to remedy this by effectively cumulating the error result from the 'P' action to increase the correction factor & finally the derivative tuning attempted to minimize this overshoot by slowing the correction factor applied as the target is approached.

In the 2nd & 3rd contribution, controller design using the multirate output feedback concepts (*periodic output feedback fast output sampling feedback*) was carried out & this developed POF/FOS controller was being used to control the various parameters of the 1-link & 2-link flexible manipulator using the POF/FOS concepts (*a type of multi-rate output feedback controller, i.e., sampling @ 2 different rates, viz.,* τ *&* Δ). The developed POF/FOS controller was put in loop with the plant (*1-link or 2-link flexible manipulator*) and the control strategy developed was tested for its effectiveness by running the developed code/model & observing the simulation results with and without the control to prove the control effectiveness.

Research was conducted on the set-point control, end-point displacements, control of the joints of the tip-1 & tip-2 of the single link and double link flexible robotic manipulator using POF/FOS concepts. CT state space model of the 1 & 2 link system was developed. The CT state space model was discretized @ different rates and the discretized systems was obtained for developing the controller. POF/FOS controllers were designed for the flexible manipulators to control the joint 1 of link 1 of 1-link manipulator & the joints 1 & 2 of – the 2 link flexible manipulator along with the tip of the end effector.

Through the simulation results, it is inferred that when the plant is placed with this designed POF/FOS controller, the plant performs well, the output tip displacements reaches the set-point quickly in the form of the closed loop response performing much better than the open loop response. From the responses of the double link flexible manipulator, it is observed that the output displacement of the link-1 is more compared to that of the 2-link case as it has to drive the link-1 & 2 plus the motor-2 weight also along with the end-effector payload, which is housed @ the shoulder level point.

A very small magnitude of control input u is required to control the actuator as it is moved away from the base in a 2-link case as a result of which less effort has to be put by the POF/FOS controller (*since the link-2 motor has to take care of only the link-2, whereas the link-1 motor has to bear both the link-1 & link-2 motor*). The magnitude of the impulse response (*closed loop*) of both the continuous and the fictitious-lifted system (*intermediate, i.e., with the open loop injection gain G – POF case or the SFB gain F – FOS case*) is less compared to their open loop counterparts.

The close loop response characteristics with the gains, viz., G and **K** (POF) or F and **L** (FOS) are also the best & settle out quickly. Thus, the observations are made with and without the controller to show the control effect. From the POF/FOS simulations, it was observed that without control the transient response was unsatisfactory, takes more time to settle and with control, the closed loop response is. The work is carried out for both the step & sinusoidal excitations.

The designed POF/FOS controller requires constant gains and hence may be easier to implement it in real time application. The controllability & observability index obtained is 2 or 4 for the SISO or MIMO models, so that all the states are controllable & observable. One excellent advantage of the proposed POF/FOS methodology developed is that the computation time required for processing & getting the output is just within 3-4 seconds, which shows the advantage of our proposed method over the others [91] & [97]. One advantage of the 2-link mechanisms is that it can cover a greater area of the workspace & the tip or the end-effector can be tracked in a bigger dimension, i.e., R^2 area as the flexible manipulator works in the planar environment ($x-y$ place).

Coming to the final contribution, i.e., in the design of controllers using the discrete sliding mode using output samples, the simulation results were observed for different sets of parameters of q, τ, N, Δ, ε & the switching surfaces $s(k)$ along with the control inputs $u(k)$ were computed. It was also seen that the responses of the states x_1, x_2 converges to zero (0) from a given initial condition $x(0)$. Also, the phase plots for the flexible system showed that the system states exhibiting the quasi-sliding motion. Another response plot showed that the switching function decreases towards zero from the initial value and stay within a small band in the neighbourhood of the switching line.

It was also observed for a 2-link case, the control effort needed for the link-1 is more than that of the link-2 as the base actuator has to control the weight of the link-1 + shoulder motor weight + the weight of the link-2 along with its end-effector payload mass, which can be justified from the control effort required for positioning diagrams in the simulation results. Further, it can also be observed that the DSM control law can be directly obtained in terms of the output samples y and immediate past control function u. The different parameters that were observed from the simulation results were – the state trajectory, phase plots, output tip displacements, system states, switching function, chattering effects & the control input.

The research work done is compared with some of the works of the yesteryears [91] & [97]. Further, one main advantage of this DSMC methodology using the concepts of output samples is that the algorithm neither needs the states of the system for feedback purpose nor for the switching function evaluation & therefore can be easily implementable in real time applications as the gains are piecewise constants.

Some of the advantages of this research work are – this has got wide applications in the field of flexible industrial robotic manipulation, smart intelligent systems, space robotics, bio-medical engineering, structural health monitoring, flexible robotics, development of light weight systems, nano-robotics, etc. In practice, flexible manipulators with two or more than two links are more desirable as such manipulators have more degrees of freedom (DOF) in movement & cover a greater work surface area.

7.2 Scope for the Future Work

Due to the experience gained during the course of this research work which was taken up by us, many research problems have come up which could be solved upon or taken up by the future researchers. Some of the future works that could be thought of to carry out further research in this exciting field of research by future researchers could be summarized as follows.

1. The simulation results presented in this research work could be experimentally verified by carrying out some real time experiments with controller kits & hardware such as the DSPACE kits, PSPICE kits, VHDL kits, FPGA kits, TI & VI kits, Power Electronics kits, etc, i.e., hardware implementation could be thought of using different types of microcontroller kits, DSP kits, dSPACE kits, NI LabVIEW kits, Raspberry pi, Python Programming, FPGA implementation, implementation using Xilinx-HDL, etc… and validating the simulation results with the hardware results.
2. Control of 1-link & 2-link flexible robotic manipulator using Quantitative Feedback Theory (QFT) scheme for controlling the set-points (position) & the vibrations of the actuators can be thought of.
3. The research work carried out could be further extended to control multi-chain-link flexible robotic manipulators.
4. Artificial Neural Networks & Convolution neural networks could also be thought of for the control purposes.
5. Hybrid algorithms can be used for better accuracy & results.
6. Development of a single controller to control both the links.
7. Usage of different types of control strategies to perform the same task as carried out in this thesis (mentioned in the introductory part).
8. Development of an accurate dynamic model could be developed by incorporating several effects such as gravity, payload rotary inertia and friction into the dynamic model of the system so that the inclusion of these parameters could provide a closer representation of the actual system.
9. Development of robust feedback control using encoders, accelerometers or gyros.

www.ingramcontent.com/pod-product-compliance
Lightning Source LLC
Chambersburg PA
CBHW050436010526
44118CB00013B/1557